Reliability and Maintainability Guideline for Manufacturing Machinery and Equipment, Second Edition

M-110.2

Published by:

Society of Automotive Engineers, Inc.
400 Commonwealth Drive
Warrendale, PA 15096-0001
U.S.A.
Phone: (724) 776-4841
Fax: (724) 776-5760

and

National Center for Manufacturing Sciences, Inc.
3025 Boardwalk
Ann Arbor, MI 48108-3266
U.S.A.
Phone: (734) 995-0300
Fax: (734) 995-4004
August 1999

Copyright ©1999 National Center for Manufacturing Sciences, Inc. and the Society of Automotive Engineers

Library of Congress Catalog Card Number: 99-64561
ISBN 0-7680-0473-X

All rights reserved. Printed in the United States of America.
First printing.

This document is protected under both the U.S. Copyright Act and applicable state trade secret laws and may not be used or disseminated without the express written permission of the NCMS or SAE. Neither the NCMS or SAE, members of NCMS or SAE, nor any person acting on behalf of them:

- makes any warranty or representation, express or implied, with respect to the accuracy, completeness or usefulness of the information contained in this document, or that the use of any information, apparatus, method, or process disclosed in this report may not infringe privately owned rights, or
- assumes any liability with respect to the use of, or from damages resulting from the use of, any information, apparatus, method, or process disclosed in this document.

SAE Order No. M-110.2

Preface

The first edition of *Reliability and Maintainability Guideline for Manufacturing Machinery and Equipment* was copyrighted in 1993 by the National Center for Manufacturing Sciences (NCMS) and distributed by the Society of Automotive Engineers (SAE) and NCMS as SAE Order No. M-110. It was well received, widely used, and was believed by the machinery and equipment industry to have contributed to improvements in equipment performance. The first edition had been written with the expectation that it would be a dynamic document, consistent with the changing requirements of the manufacturing machinery and equipment marketplace. Since the guideline's initial publication in 1993, the passage of time and the extent of industry usage marked the need for a revision. In addition, the need for a second edition was stimulated by changes to the Tooling & Equipment Supplement to QS-9000 along with a broadened application of reliability and maintainability (R&M) by the transportation industry.

Undertaken as a joint effort of SAE and NCMS, this second edition, SAE Order No. M-110.2, was composed by a consortium team representing three subgroups: user companies that use machinery and equipment in their manufacturing process; supplier companies who design and manufacture the machinery and equipment; and associations, not-for-profit organizations that represent educational, research, membership, trade group, and industry interests in R&M.

This consortium held the basic tenets that the second edition should:

- Remain a guideline
- Convey common language for R&M
- Encourage user and supplier partnerships
- Present an organized framework for an R&M program
- Retain focus on the transportation industry
- Equalize emphasis on Maintainability with Reliability.

Secondary tenets were to:

- Improve the readability; convey the message more clearly and consistently than the first edition
- Simplify the checklists
- Make appendices more informational
- Add technical references.

The second edition adopts a three-section style:

Section I is composed of two chapters forming an introduction to the implementation of R&M.

Section II is composed of five chapters dealing with R&M in each of the five life cycle phases of machinery and equipment. Each chapter has the following sections:

- **Introduction** — summarizes *what* R&M activities may occur during that phase and *why*

- **User's Role** and **Supplier's Role** — explain *who* may perform *what* R&M activities during the phase

- **Recommended Practices** —describes *what* R&M tools and techniques are associated with the phase and *how* they may be used.

Section III is composed of supporting documentation, appendices, and reference materials.

This second edition remains focused on conveying to personnel at all levels the fundamentals of reliability and maintainability practices that may be used during the life cycle phases of manufacturing and equipment. The guideline retains as its central theme the continuous improvement of machinery and equipment, ultimately for the improvement of succeeding designs through the use of cooperative R&M practices. Users and suppliers are advised to select and use only those practices and measures that are appropriate for the actual type and intended purpose of the machinery and equipment.

Acknowledgments

The following companies participated in the project to revise the *Reliability and Maintainability Guideline for Manufacturing Machinery and Equipment*. This edition was prepared with their support.

Automotive Industry Action Group
Carboloy Inc.
Cincinnati Machine, a UNOVA Company
Cobra Patterns & Models Inc.
Comau North America
Creatron Corporation
Cross Hüller Special Machine Systems, a Thyssen Production Systems company
Cutler-Hammer
DaimlerChrysler Corporation
DRM Technologies
Durr Automation
Eaton Corporation
Excel Partnership Inc.
Ford Motor Company
General Motors Corporation
Gilman – Giddings & Lewis, Inc.
Industrial Technology Institute
Ingersoll Milling Co.
Ingersoll-Rand
Kennametal Inc.
Kuka Welding Systems & Robot Corporation
Lamb Technicon, a UNOVA Company
Landis Gardner, a UNOVA Company
Management Resources International, Inc.
Max Systems
National Center for Manufacturing Sciences
Omnex
Parker Hannifin Corporation
Piviola Tool
Progressive Tool & Industries Company (PICO)
R&M Associates
Rockwell Automation
RTW Greenfield Kennametal
Sandvik Coromant
Schneider Electric (Modicon)
Society of Automotive Engineers International
Square D Company
The Association for Manufacturing Technology
TRW
Valenite, Inc.
Vogel Lubrication, Inc.
WCA, Inc.
Weldun International
Western Michigan University

Several representatives from these companies participated in a consortium steering committee and in preparing this text. Their contributions were essential to the successful completion of the Guideline. Their commitment to this effort was exceptional.

Dave Rogers and David Janas	DaimlerChrysler Corporation
Gary Givens and Paul Helmstetter	Ford Motor Company
Donna Sajdak and Buck Steele	General Motors Corporation
Stan Meitzner	Cross Hüller Special Machine Systems
Ed Rivard	Lamb Technicon
Aron Brall	Landis Gardner
Mark Morris	Management Resources International, Inc. (formerly with PICO)
Ben Tower and Victor Malinasky	Progressive Tool & Industries Company
Paul Warndorf	The Association For Manufacturing Technology
Constance J.S. Philips	National Center for Manufacturing Sciences
Jack Pokrzywa	Society for Automotive Engineers International

Initial collection of revision suggestions was facilitated by Tim Faricy, R&M Associates. Project management was provided by Constance J.S. Philips, NCMS. Editorial support during the final preparation for publication was provided by Martha Swidersky, NCMS.

Table of Contents

Section I: Introduction to R&M and Its Implementation

Reliability and Maintainability (R&M) are vital characteristics of machinery and equipment. Improved reliability and maintainability lead to lower life cycle costs.

Successful R&M implementation requires commitment, strategy, and action. It also requires planned activities during each phase of the machinery and equipment life cycle process.

Table of Contents (continued)

Section II: R&M and the Life Cycle Process

During the Concept and Proposal phase, the performance requirements and R&M qualitative and quantitative requirements for manufacturing machinery and equipment are established, preliminary design concepts are considered, and suppliers' proposals addressing how the requirements specified by the user will be met are evaluated. This process leads to the selection of an equipment supplier.

The Design and Development phase focuses on machinery and equipment design and verification of the capability of the evolving design to meet the R&M requirements specified in the Concept and Proposal phase.

Table of Contents (continued)

During the Build and Install phase, care must be taken to maintain inherent R&M, those R&M capabilities designed into the equipment. The supplier's manufacturing process variables affecting inherent R&M should be identified and targeted for control. Installation-related anomalies should be addressed prior to machine start-up to reduce the effect of compounding failures.

In the Operation and Support phase, the user is expected to implement a system of R&M data collection, analysis, and feedback. The supplier uses feedback to improve the R&M of existing as well as new machinery and equipment designs.

Table of Contents (continued)

The Conversion or Decommission phase marks the end of the expected life of a piece of machinery or equipment. When either conversion or decommission occurs, feedback from the user plant should be considered for R&M growth and continuous improvement in future generations of machinery.

Section III: Supporting Documentation

List of Figures and Tables

Chapter One

Introduction to Reliability and Maintainability

Reliability and Maintainability (R&M) are vital characteristics of machinery and equipment. Improved reliability and maintainability lead to lower life cycle costs.

Introduction

This guideline provides reliability and maintainability fundamentals, for personnel at all levels (users and suppliers), that are commonly used in the life cycle phases of manufacturing machinery and equipment to support up-front engineering in the design process. It describes recommended R&M practices, including tools and techniques for both users and suppliers of machinery and equipment.[1]

The purpose of reliability improvement is to increase manufacturing productivity and throughput. Designing reliability into machinery and equipment directs the improvement effort toward reducing the frequency of failure, increasing the durable life, or both. As failures occur less often, the need for corrective maintenance is reduced.

A purpose of maintainability improvement is to design machinery and equipment that can be repaired quickly and safely to reduce downtime. Another is to identify planned maintenance actions that improve the reliability and longevity of equipment.

Key Definitions[2]

- **Reliability** is the probability that machinery and equipment can perform continuously for a specified interval of time without failure, when operating under stated conditions. **Increased reliability** implies less failure of the machinery and equipment and, consequently, less downtime and loss of production.
- **Maintainability** is a characteristic of design, installation, and operation of machinery and equipment. Maintainability is usually expressed as the probability that a machine can be retained in or restored to specified operable condition within a specified interval of time, when maintenance is performed in accordance with prescribed procedures.
- **Availability** is a percentage measure of the degree to which machinery and equipment is in an operable and committable state at the point in time when it is needed.
- **Life Cycle Cost (LCC)** is the total cost of ownership of machinery and equipment, including its cost of acquisition, operation, maintenance, conversion, and/or decommission.

---

[1] The terms "manufacturing machinery and equipment," "equipment," and "machinery" are used interchangeably in this guideline. The term "supplier" includes the original manufacturer and sub-suppliers of components or subsystems that are integrated into machinery and equipment end products.

[2] Other terms used in this guideline are defined and referenced in the Glossary.

Benefits of Reliability and Maintainability

Reliability and maintainability are vital characteristics of manufacturing machinery and equipment that enable manufacturers to be foremost competitors worldwide. Effective production planning depends on a manufacturing process that yields high-quality parts at a specific rate without interruption. R&M characteristics that are predictable for machinery and equipment used in manufacturing processes are key factors in attaining and maintaining efficient production and in utilizing advanced manufacturing practices. Improved R&M leads to lower life cycle costs that are necessary to maintain a competitive edge.

This document supports these objectives by describing R&M practices and providing guidance on when to apply them, from the conceptualization of new machinery and equipment through each of the subsequent life cycle phases of that equipment. Successful implementation of R&M practices requires cooperative efforts from both user and supplier. Neither participant in the process can accomplish the objectives alone.

Manufacturing machinery and equipment that have high availability offer the means for producing high-quality products consistently at lower cost and at higher output levels. Table 1 shows how improved R&M benefits both the machinery and equipment user and supplier.

Table 1. Improved R&M User/Supplier Benefits

User Benefits	Supplier Benefits
• Higher availability of machinery and equipment	• Reduced warranty costs
• Unscheduled downtime reduced/eliminated	• Reduced build, install, and support costs
• Reduced maintenance costs	• Improved user relations
• Stabilized work schedule	• Higher user satisfaction
• Lower equipment LCC	• Improved status in the marketplace
• Improved profitability	• A competitive edge in the marketplace
• Increased employee satisfaction	• Increased employee satisfaction
• Lower overall cost of production	• Increased understanding of product performance
• More consistent part/product quality	• Feedback for improved designs
• Less need for in-process inventory to cover downtime	
• Improved safety	
• Improved resource utilization	

Safety

Accidents and injury to manufacturing personnel can occur during machinery operation or during repair and maintenance. The benefits of an R&M program as it relates to safety are derived from *designing for reliability* to reduce the frequency of breakdowns, and from *designing for maintainability* to facilitate quick and safe repair and maintenance. Throughout any R&M program, consideration must always be given to safety. Benefits of an improved design must not be allowed to compromise the ability of machinery and equipment to be operated safely, and to be maintained without risk to personnel. A detailed treatment of safety is beyond the scope of this guideline and we refer the reader to safety professionals and government regulatory agencies.

Life Cycle Cost

Life Cycle Cost (LCC) refers to the total cost of a system during its life cycle (See Appendix A). LCC is the sum of the costs incurred in all phases of the life of a machine. Figure 1 illustrates four of the five life cycle phases for machinery and equipment. The costs incurred during the first three phases, Concept and Proposal, Design and Development, and Build and Install are considered non-recurring costs. Costs incurred during the fourth phase, Operation and Support, are considered recurring costs, as they reoccur throughout the operational life of the machinery and equipment. The Operation and Support phase typically contributes the majority of costs represented in LCC. Figure 1 graphically shows the relative proportion of costs contributing to LCC during the first four phases. The costs associated with the fifth phase, Conversion or Decommission, which are also considered non-recurring costs, vary widely based on the circumstances and are not addressed here. Percentages shown are for illustrative purposes only. Actual experience varies by industry.

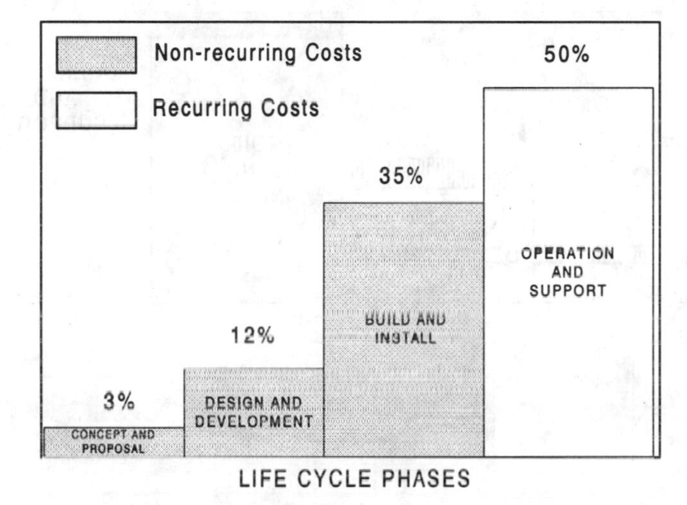

Figure 1. Life Cycle Cost During First Four Phases

Designed-in Reliability and Maintainability To Reduce Life Cycle Cost

One of the primary drivers for applying R&M to machinery and equipment during the early life cycle phases, specifically the Concept and Proposal and the Design and Development phases, is to reduce operation and support costs. The goals of any R&M effort are to improve the *designed-in reliability* and *designed-in maintainability* of the equipment. Reliability and maintainability have a direct relationship with lower Life Cycle Costs. For example, by using R&M principles and practices in the design phase to minimize stress (electrical, mechanical, thermal, etc.), the equipment will be less prone to failure during operation, thereby contributing fewer operation and support costs. As operation and support costs account for the majority of LCC, LCC will decrease. The key, then, to successful realization of R&M gains as well as the reduction of LCC is participation in R&M centered activities in these early life cycle phases.

Figure 2 consists of two graphs that are linked by the timing of the life cycle phases. The top graph shows how a slight percentage increase in spending to implement R&M during the early life cycle phases can reduce the costs incurred in the later phases and thereby dramatically impact LCC. The bottom graph illustrates the percentage range of LCC that may be predetermined by the time each of three design reviews occur. The work performed in the Concept and Proposal and the Design and Development phases can result in the predetermination of as much as 95% of LCC. This percentage range is shown as the shaded area. Once newly designed equipment reaches the Build and Install phase, the opportunity to significantly impact LCC will have passed.

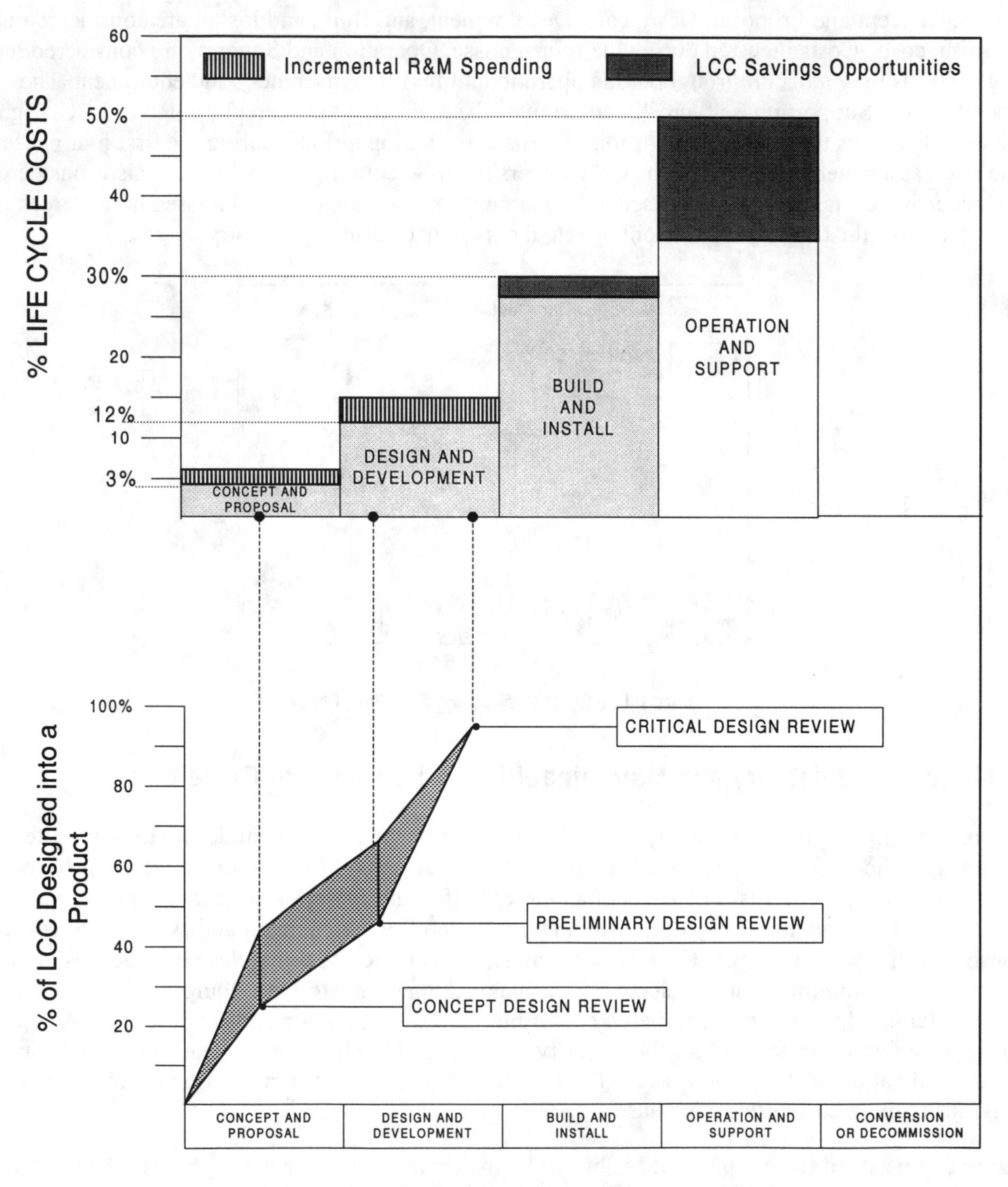

Figure 2. R&M Impact on LCC

Chapter Two
Implementing R&M Through the Life Cycle Process

Successful R&M implementation requires commitment, strategy, and action. It also requires planned activities during each phase of the machinery and equipment life cycle process.

Introduction

Successful implementation of R&M depends upon structured communication between the user and the supplier throughout the machinery and equipment life cycle. The need for this communication begins at project conception and continues through all phases of the life cycle for the purpose of ensuring that equipment problems are prevented or identified, root causes determined, and corrective actions implemented. In addition, it is important that "Lessons Learned" are considered on new designs.

Management Commitment

Attainment of reasonable levels of R&M seldom occurs by chance. It requires planning, goal definition, a design philosophy, and an operational philosophy that includes analysis, assessment, and feedback for continuous improvement. Management must recognize the value of R&M and commit the personnel and resources necessary to attain the goal. Without such a commitment, the probability of attaining R&M goals throughout the life cycle is low. Successful attainment of the R&M quantitative and qualitative goals requires a trained and focused team effort involving appropriate functions of both the user and supplier.

Life Cycle Phase Program Management Process

There are five phases in the machinery and equipment life cycle process. The life cycle starts with the Concept and Proposal phase, and proceeds through the Conversion or Decommission phase. This phased process is appropriate for any machinery and equipment development program. Figure 3 shows the five phases of the machinery and equipment life cycle process. Each of the five phases is discussed more fully below. Manufacturing machinery and equipment suppliers need to use methods such as those stressed in this guideline and in other more detailed reliability and maintainability sources to ensure that R&M requirements will be met as the machinery and equipment progress through these life cycle phases.

Concept and Proposal Phase	Design and Development Phase	Build and Install Phase	Operation and Support Phase	Conversion or Decommission Phase

Figure 3. Five Phases of Manufacturing Machinery and Equipment Life Cycle

Concept and Proposal Phase

This first phase includes research, limited development or design, and written proposals leading to the selection of an equipment supplier. During this phase, the user and the supplier work together to establish meaningful system requirements, clearly stated requirements for future monitoring, and the division of responsibilities for collecting, analyzing, and reporting of data. The user team can include machine operators, maintenance personnel, and engineering personnel. The supplier team can include engineering, service and assembly floor personnel, and subcontractors. Machinery mission and environmental requirements are defined during this phase, as are safety issues, goals for R&M, and LCC objectives. A concept design review may occur during the Concept and Proposal phase. Chapter Three, User and Supplier R&M Activities in the Concept and Proposal Phase, describes design reviews. For suppliers of standard commercial equipment, where R&M requirements are set principally through a company's internal processes, goals for R&M and LCC should be instituted in these internal processes.

Simultaneous or concurrent engineering can be introduced in either the Concept and Proposal phase or the Design and Development phase (preferably, in this earlier phase), depending on the particular situation and type of manufacturing machinery or equipment.

Design and Development Phase

The Design and Development phase addresses the reliability, maintainability, and performance issues that arose during the Concept and Proposal phase, such as safety, ergonomics, accessibility, etc., for design into the system. R&M requirements are formalized. Components and component suppliers should be selected based on criteria that include predictive R&M statistics.

Design reviews ensure that the planned design is likely to meet all requirements in a cost-effective manner, considering all variables and constraints while giving special attention to maintainability. Typically, two design reviews occur during the Design and Development phase. A preliminary design review precedes commitment to a design approach. A critical design review determines overall readiness for production prior to release of drawings to the manufacturing function.

Additional design meetings may be held regularly to ensure clear communication between the user and machinery and equipment suppliers. User machine operators, maintenance personnel, and product and process engineers should participate in these meetings so that all concerned understand the design intent.

At this phase, the R&M plan can include suitable verification and validation test plans, when agreed to by both the user and the supplier, to demonstrate compliance to R&M requirements. The design must also provide for disassembly and re-assembly of the machinery and equipment in the user's plant.

Build and Install Phase

During the manufacturing and assembly of the machine, the achievement of R&M requirements needs to be monitored. Issues that could affect R&M are communicated back to the design engineers to ensure that any redesign includes R&M improvements. Manufacturing process variables affecting R&M should be identified during this phase and targeted for control. Events that occur during this phase requiring specific attention include:

- The development of maintenance procedures (A user representative needs to be involved in this process.)
- Training that starts at this phase and continues into the next phase
- Machine acceptance tests, as previously agreed, that may be performed prior to disassembly and installation
- R&M data collection during machine acceptance testing
- The identification of crucial assembly processes prior to disassembly, transportation to the user's facility, and re-assembly at installation
- The documentation and elimination of any infant mortality (premature) failures that occur during the initial start-up.

Operation and Support Phase

The machinery and equipment, delivered and installed at the user location, will be fully operational in this phase. Data collection and feedback now become very important. Data collection methods and procedures should be in place and agreed upon by both parties. Access to machinery and equipment maintenance data and R&M databases can facilitate the success of an R&M initiative. Information collected during this phase is also used to facilitate R&M growth and continuous improvement for future designs, or for the current design if contractual requirements are not met. During this phase, preventive and predictive maintenance ought to be performed regularly.

Conversion or Decommission Phase

This phase marks the end of the operational life of a machine. When an increasing failure rate results in increasingly expensive maintenance, the user may chose to convert or decommission a machine. When a machine is still useful, either in its current production installation or in another application, the user may chose to convert it, by having it reworked, retrofitted, rebuilt, or remanufactured. When the user deems that a machine is to be decommissioned, it is removed and sold, scrapped, stored, or transferred to another location. When either conversion or decommission occurs, feedback from the user plant should be considered for future design decisions, as the information may be useful for R&M growth and continuous improvement in future generations of machinery.

Summary of Phases

Figure 4 summarizes what should be performed in each of the five phases.

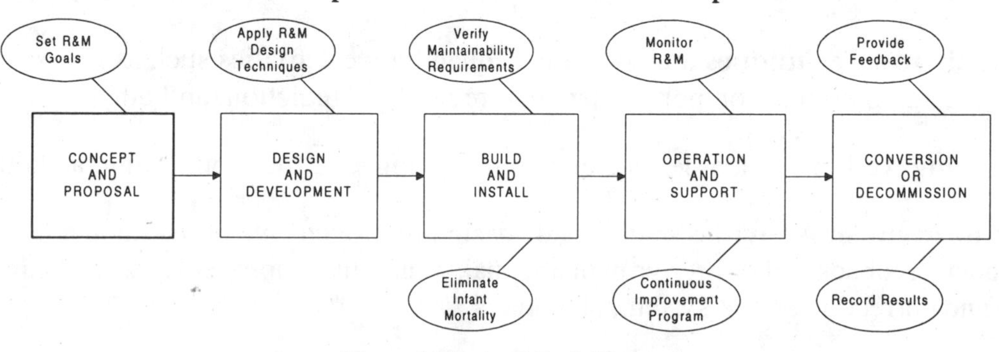

Figure 4. Goals of Each Phase

Keys to Successful Implementation of R&M

This guideline identifies seven keys to the successful implementation of R&M throughout the life cycle phases. They are: R&M planning, quantitative requirements, design assurance activities, user feedback, R&M analytical techniques, R&M testing activities, and R&M improvement activities. They are shown in Figure 5 and are discussed below.

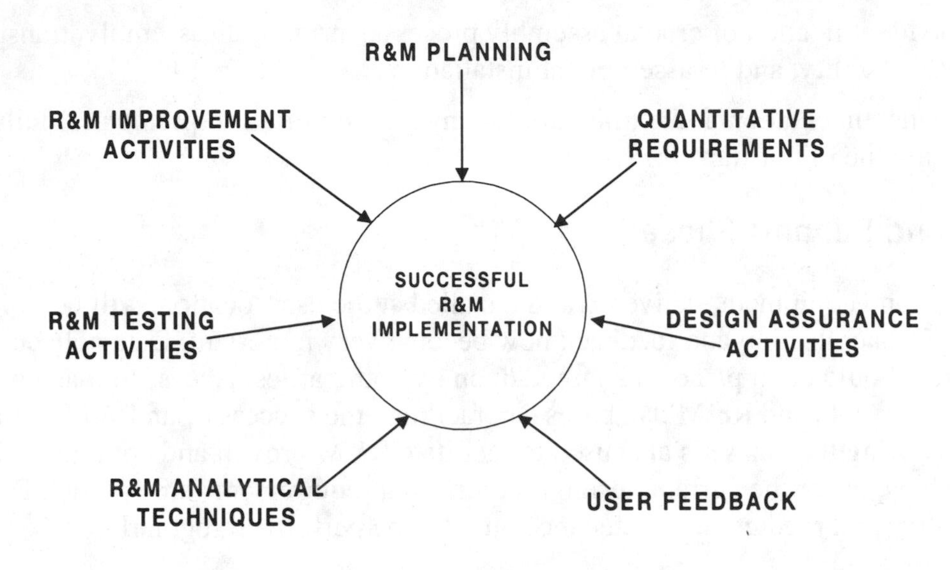

Figure 5. Keys to Successful R&M Implementation

1. **R&M Planning** is a structured approach to R&M that requires suppliers and users to focus on how R&M requirements will be met—resulting in a project-specific R&M plan. Such a plan might list agreed-upon activities, responsibilities, and milestone dates.

2. **Quantitative Requirements** are figures of merit defined in the Concept and Proposal phase.

3. **Design Assurance Activities** include qualitative activities identified in a project-specific R&M plan. Such activities might include structured design reviews, failure mode and effects analysis, component selection, maintainability assessment, and others.

4. **User Feedback** includes equipment performance data, maintenance logs, lessons learned, lists of plant concerns, and others.

5. **R&M Analytical Techniques** include various quantitative activities such as stress analysis, setting of design margins, component derating, reliability prediction, and others.

6. **R&M Testing Activities** include accelerated life testing, qualification testing, and others.

7. **R&M Improvement Activities** may include design improvements based upon equipment performance feedback, reliability growth and maintainability improvement, failure reporting, analysis and corrective action system, and others.

Sample R&M Application in a Design Process

R&M for machinery and equipment must be addressed in a systematic manner. Figure 6 provides an example of a systematic process flow for the application of R&M. Graphic depictions can vary based on industry and the circumstances pertinent to the machinery and equipment.

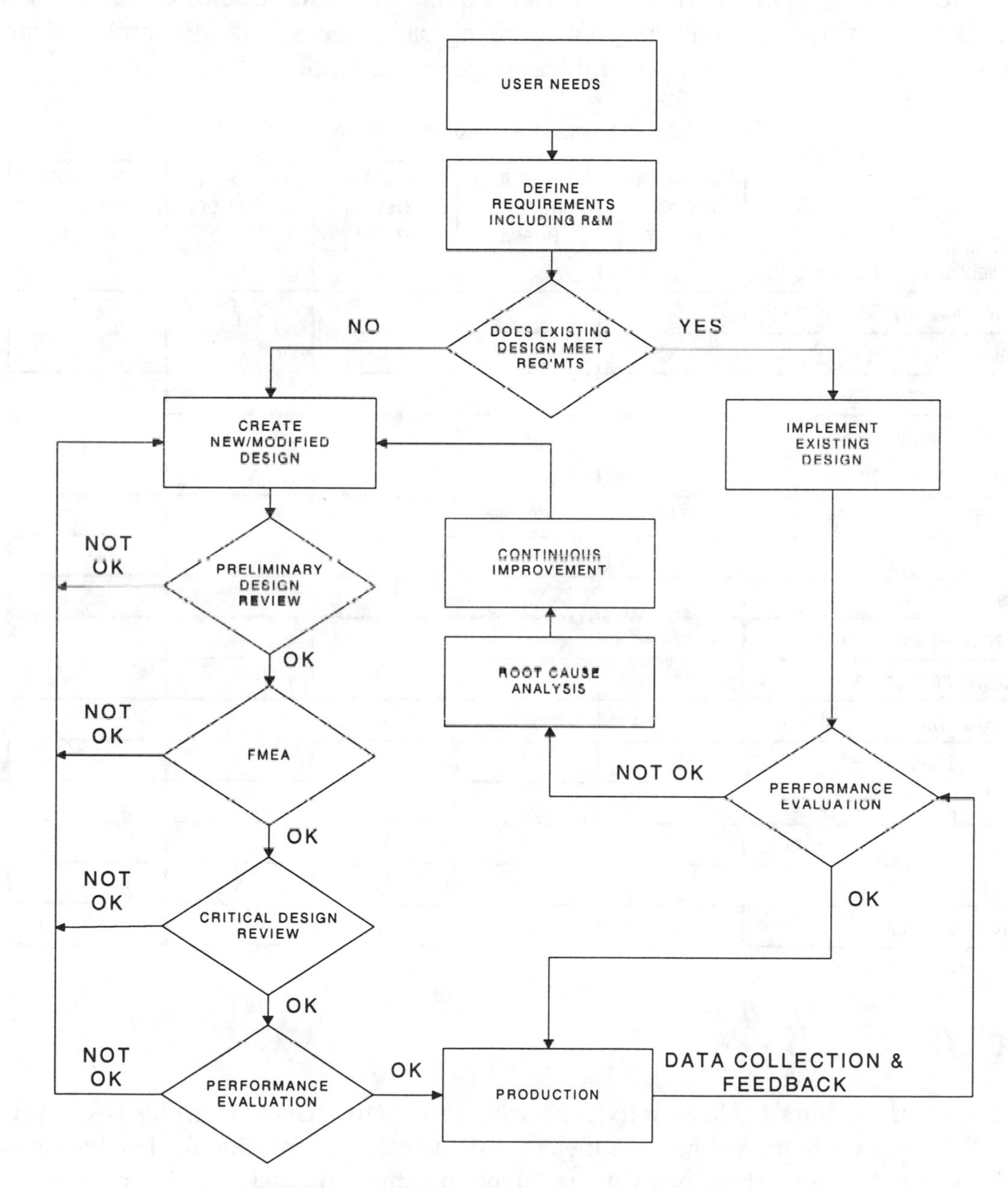

Figure 6. Sample R&M Application in a Design Process

Tailoring R&M Activities Over the Life Cycle Phases

R&M activities planned over the machinery and equipment life cycle phases are a function of both user machinery requirements and the capabilities of the selected supplier. Not every activity is appropriate for a given contract. Tailoring of the machinery requirements by the user allows cost effective application of R&M program elements. Table 2 provides an example of an R&M program matrix. The intent of a program matrix is to provide guidance as to when selected activities might be appropriate. Contractual requirements may dictate changes in responsibility. Examples of program matrices tailored for differing contractual circumstances can be found in Appendix B.

Table 2. Sample R&M Program Matrix

	Concept and Proposal Phase	Design and Development Phase	Build and Install Phase	Operation and Support Phase	Conversion or Decommission Phase
Reliability Requirements	X				
Maintainability Requirements	X				
Failure Definition	X				
Environment/Usage	X				
Design Margin		X			
Maintainability Design		X			
Reliability Predictions		X			
FMEA/FTA		X			
Design Reviews		X			
Machinery Parts Selection		X			
Tolerance Studies		X			
Stress Analysis		X			
R Qualification Testing			X		
R Acceptance Testing			X		
R Growth/M Improvement		X	X	X	
Failure Reporting			X	X	
Data Feedback	X	X	X	X	X

SAMPLE

R&M Activity Matrix

Appendix C contains a sample R&M Activity Matrix. This matrix lists the various R&M activities occurring in each life cycle phase and identifies who is typically responsible for leading or supporting the activity. Additionally, the organizational function commonly assigned that responsibility is identified. Users and suppliers can use this type of matrix as a planning tool.

Chapter Three

R&M Activities During the Concept and Proposal Phase

During the Concept and Proposal phase, the performance requirements and R&M qualitative and quantitative requirements for manufacturing machinery and equipment are established, preliminary design concepts are considered, and suppliers' proposals addressing how the requirements specified by the user will be met are evaluated. This process leads to the selection of an equipment supplier.

Introduction

The Concept and Proposal phase includes research, limited development or design, and written proposals leading to the selection of an equipment supplier. During this phase, the user and the supplier work together to establish machinery and equipment specifications and Reliability and Maintainability (R&M) requirements. As part of this effort, machinery mission and environmental requirements are defined during this phase, as are safety, goals for R&M, and LCC objectives.

A cross-functional team approach to the activities undertaken in this phase is recommended. Such a cross-functional user team can include personnel from production operations, maintenance, and engineering. A supplier team can include engineering, service and assembly floor personnel, and subcontractors. Simultaneous or concurrent engineering can be introduced in the Concept and Proposal phase, depending on the particular situation and type of manufacturing machinery or equipment.

During this phase, the supplier and the user may negotiate meaningful R&M goals, clearly state the requirements for future monitoring, and mutually determine the responsibilities for collecting, analyzing, and reporting of data. During this phase, the user and supplier often communicate and discuss proposed concepts. (Refer to "Designed-In Reliability and Maintainability To Reduce Life Cycle Costs" in Chapter One for more information.) A concept design review may occur also. The desired outcome of this phase is the successful preparation of an R&M plan by the supplier and acceptance concluding with a contract or purchase order issued by the user to the supplier.

User's Role in the Concept and Proposal Phase

During the Concept and Proposal phase, the user identifies the R&M goals to be accomplished for manufacturing machinery and equipment purchases and conveys those goals to potential suppliers through R&M specifications, such as tailored program matrices and R&M program planning worksheets. See Appendix B for tailored program matrices and Appendix D for an R&M Program Planning Worksheet. Users also identify the machinery use, duty cycle, throughput, and environment in which the machinery and equipment will be required to perform as well as define specific R&M and data requirements, and the reliability testing that should occur to qualify the machinery and equipment. LCC objectives may be established for use in the user's acquisition process. In addition, an

LCC projection methodology or model may be created to evaluate projected LCC of various systems under consideration. See Appendix A for additional LCC information.

Supplier's Role in the Concept and Proposal Phase

During the Concept and Proposal phase and prior to contract award, the potential supplier needs to gain a full understanding of the evolving R&M requirements. In the written proposal and during proposal-related presentations, the supplier should be prepared to address how the R&M requirements will be met. The result may be an R&M plan submitted to the user that addresses R&M activities, responsibilities, deliverables, and milestones to be implemented following contract award in the supplier's facility during the Design and Development and Build and Install phases. Additionally, the supplier will define the scope and number of design reviews, held both internally and with the user during these phases, and include this information in the R&M plan submitted to the user.

Recommended Practices in the Concept and Proposal Phase

Table 3 lists the prevailing tools and techniques that are recommended R&M practices for users and/or suppliers during the Concept and Proposal phase. Each is discussed further in the text below.

Table 3. Recommended R&M Practices in the Concept and Proposal Phase

Tools and Techniques	User	Supplier
Planning	X	X
R&M Plan		X
Lessons Learned	X	X
Specify Reliability & Maintainability Requirements	X	
Machinery Use	X	
Duty Cycle	X	
Machinery Environment	X	
Continuous Improvement Monitoring	X	X
Life in Terms of Throughput	X	
Data Collection	X	X
R&M Program Matrix	X	
R&M Program Planning Worksheet	X	
Design Reviews	X	X

See Appendix E for information on other tools and techniques for reliability and maintainability.

Planning

R&M planning determines how the R&M goals necessary to meet operational requirements for manufacturing machinery and equipment will be achieved. To perform R&M planning, it is necessary to have a process for *designing for R&M*, coupled with a process for ensuring that *inherent R&M* is achieved during the Design and Development, Build and Install, and Operation and Support life cycle phases of manufacturing machinery and equipment. The resulting plan identifies what should be done, who should do it, and when it should be done.

The thoughtful and thorough development and communication of machinery and equipment specifications and R&M requirements by the user are the most fundamental elements of ensuring achievement of R&M goals. All subsequent design decisions, manufacturing activities, and service systems are based on those needs, in whatever form they take. Specifications, in this context, refer to descriptions of machinery and equipment performance requirements, which prescribe what the machinery is supposed to be and do, not how to design, build, or test it.

To accomplish the planning of R&M requirements, the user should organize an internal cross-functional team, which may include personnel from product design engineering and process engineering, representatives from purchasing, manufacturing, quality control, finance, maintenance skilled trades, production operations, and additionally, may include representatives from the selected supplier.

The cross-functional team then develops a written plan outlining the specific R&M activities to be undertaken to ensure machinery and equipment reliability and maintainability performance. R&M program matrices and worksheets, such as those found in Appendices B and E, are developed to help users and suppliers structure and facilitate their planning efforts to achieve the desired R&M goals. It is important to understand that if machinery and equipment are not *designed for R&M*, it may be cost prohibitive to significantly improve R&M later.

R&M Plan

The supplier may prepare an R&M plan as part of the written proposal to the user. An R&M plan addresses the reliability and maintainability activities that are required to achieve the R&M requirements specified in the user's tailored R&M program matrices and R&M program planning worksheets. It should set forth the supplier's implementation strategy for performance of R&M activities, design assurance, design reviews, reliability testing and assessment, and the assignment of responsibilities, deliverables, and milestone schedules to be implemented following contract award. When required by contract, an R&M plan would also address the supplier's role in support of the user's continuous improvement activities that occur throughout the machinery and equipment life cycle, and additionally, the process by which failure issues are to be resolved.

Lessons Learned

To avoid the repetition of past mistakes, to recover past solutions, and to improve machinery and equipment designs, it is important to address the lessons learned from previous applications and to document lessons learned for use on future applications. Information that can be gleaned from this activity includes desirable features, past problems, changes implemented during build and install, and those changes incorporated in the machinery and equipment during plant operation by either the supplier or user.

Specifying Reliability and Maintainability Requirements

The user defines R&M requirements to meet R&M goals in quantitative terms. Appendix F lists common Figures of Merit that may be used to specify reliability and/or maintainability requirements.

The values assigned to R&M requirements indicate the desired inherent reliability and maintainability to be designed into the machinery and equipment. While the supplier will not be held accountable for nonconformance caused by operation outside these stated requirements and the supplier's stated operating conditions, both the supplier and user need to recognize that real-world occurrences may impact actual performance. When R&M requirements are defined by using Figures of Merit, it is important that the user and supplier define failure. All downtime events are considered in the definition of failure in this guideline.

Machinery Use

Machinery suppliers and users need to be aware of how the machinery is to be used and how reliability will be impacted by this use. For example, cycle time reductions obtained by increasing speeds and feeds can reduce reliability. Similarly, increased complexity and extending the length of in-line operations can increase the failure rate and diagnostic time. Minimizing, through design, the time required to perform scheduled maintenance in terms of accessibility, inspection, and service can improve availability of the machinery and equipment.

Duty Cycle

Many machinery and equipment components are sensitive to start-up, power-up, and/or applied loads. Therefore, the anticipated duty cycles and intermittent operation of machinery and equipment components need to be understood and addressed in the design of the machinery and equipment.

Machinery Environment

Those who specify and design machinery and equipment need to fully understand the variations and extremes of environmental conditions to which the machinery and its components will be subjected. These environmental elements include heat, humidity, contamination, shock, and vibration at various locations within and around the machine. These factors influence the performance of mechanical, hydraulic, pneumatic, electrical, and electronic components (including items contained in enclosures) to different degrees. Environmental considerations are discussed further in Appendix E. The user and supplier should develop a thorough understanding of these factors early in the Concept and Proposal phase.

Continuous Improvement Monitoring

New manufacturing machinery and equipment reliability and maintainability can be improved greatly by using feedback from the user concerning any improvements made to the supplier's prior-generation machinery and equipment in operation at the user's plant. Feedback to the supplier is most effective if the user supplies it early in the Concept and Proposal phase. This information is typically provided in terms of "Lessons Learned" which include, but are not limited to the:

- Desirable features to be retained
- Types of machine problems experienced
- Corrective actions or improvements made
- Results of those improvements.

Improvements may be in the areas of operating cost, safety, ergonomics, and quality. The supplier may be aware of some of the improvements made to manufacturing machinery and equipment in the user's plant due to involvement with the user's continuous improvement activity. This type of partnership provides timely and effective data for use in incorporating improvements into the design of new manufacturing machinery and equipment.

Life in Terms of Throughput

During the Concept and Proposal phase, the user needs to provide the supplier with a clear definition of the throughput and product mix requirements over the projected life of the equipment. To avoid excessive costs associated with over-design or failure to meet R&M goals through under-design, clearly defined requirements are necessary to ensure supplier understanding so that an efficient design-for-use approach can be taken during these phases.

Data Collection

During the Concept and Proposal phase, user manufacturing engineering and maintenance, in co-operation with the supplier, need to specify if automatic data collecting systems (hardware/software) are to be designed into the manufacturing machinery and equipment. The systems may need to interface with the plant's overall R&M feedback system while providing relevant data to the supplier. Appendix G contains further information regarding data collection, tracking, and feedback. If manual data collection is to be used, the structure, format, and frequency need to be determined early, and be understood and accepted by all persons responsible for collecting and processing the data.

R&M Program Matrix

A procurement package issued for machinery and equipment intended for a specific user application may include a tailored R&M program matrix. (See Table 2, Sample R&M Program Matrix, and Appendix B for additional tailored matrices.) This program matrix will define the application-specific R&M program desired for the equipment being procured. In defining the R&M program a user considers the following: their assessment of the supplier's understanding of R&M practices, supplier capability, and the estimated amount of design engineering required to meet specified performance requirements. Tailoring a matrix for a specific purchase involves a case-by-case assessment of R&M needs.

The tailoring of R&M program requirements to the unique application being considered allows the technical activity to determine the R&M requirements that will yield the highest returns.

R&M Program Planning Worksheet

An R&M Program Planning Worksheet (see Appendix D) should accompany a tailored R&M program matrix in the procurement package. The worksheet is used to provide expanded detail relating to the specific program elements cited in the R&M program matrix. For example, the R&M program matrix may indicate that Reliability Qualification Testing is required in the Build and Install phase. The worksheet should detail which components and subassemblies are to be subjected to Reliability Qualification Testing.

Design Review

A design review is a formalized, documented, and systematic process for design management. A concept design review may be part of the Concept and Proposal phase. Through this process, the machinery and equipment supplier alone or the supplier together with the user will review all technical aspects of the evolving design, including R&M. This review typically involves the examination of drawings, sketches, engineers' notebooks, analysis results, test documentation, mockups, assemblies, and other hardware and software depictions of the evolving design. Design review objectives appropriate for each of the life cycle phases are listed in Table 4.

Table 4. Design Review Objectives

Life Cycle Phase	Review Objective
1. Concept and Proposal	*Concept Review*: Focuses on feasibility of the proposed design approach
2. Design and Development	*Preliminary Design Review:* Verifies the capability of the evolving design to meet all technical requirements *Final Design Review:* Validates that the documented design and related analyses are complete and accurate
3. Build and Install	*Build:* Addresses issues resulting from machine build and runoff testing *In-Plant Installation:* Conducts failure investigation of problem areas for continuous improvement

Not all design reviews are applicable to every machine. The supplier and the user should concur on which reviews are required for a particular machine and when they should be scheduled (where required by contract or purchase order). For large systems, the machinery user and the key component suppliers should participate in and be an integral part of the design review process.

R&M Activity Matrix

Appendix C provides an example of a matrix of responsibilities for R&M program activities during the Concept and Proposal phase. A user and/or supplier can use this matrix.

Chapter Four
R&M Activities During the Design and Development Phase

The Design and Development phase focuses on machinery and equipment design and verification of the capability of the evolving design to meet the R&M requirements specified in the Concept and Proposal phase.

Introduction

The activities in the Design and Development phase have a major impact on both the R&M and the LCCs of machinery and equipment. The user and supplier work together to mutually understand the specific activities and requirements outlined in the R&M plan, to verify that the supplier has addressed all specified activities and requirements, and to verify that R&M allocation requirements have been met. Components and component suppliers need to be selected based on criteria that include predictive R&M statistics. Manufacturing machinery and equipment suppliers should use R&M methods and techniques, such as those stressed in this guideline and as may be found in other more technically detailed sources, to ensure that R&M requirements will be met. In addition to reliability issues, the Design and Development phase deals with all issues that arose during the Concept and Proposal phase, such as safety, ergonomics, accessibility, maintainability as well as other identified issues for design into the machinery and equipment. All activities and identified issues should be monitored and progress tracked through periodic meetings and scheduled design reviews.

Design reviews in this phase ensure that the planned design is likely to meet all requirements in a cost-effective manner, considering all variables and constraints, while giving attention to reliability and maintainability. Typically, two design reviews can occur during the Design and Development phase. A preliminary design review precedes commitment to a design approach. A final design review determines overall readiness for production prior to release of drawings to the manufacturing function.

In addition, design meetings should be held regularly to ensure that clear communications between the user and machinery and equipment suppliers are maintained. It is recommended that user personnel from activities such as production operations, maintenance, and product and process engineering participate in these meetings so that all affected parties can gain an understanding of the design intent. During this phase the user and supplier together verify that Lessons Learned from past concepts and designs are addressed in the new design. This use of Lessons Learned will aid in the development of increasingly reliable and maintainable equipment designs.

User's Role in the Design and Development Phase

During this phase, the user works closely with the supplier to ensure continued, thorough understanding of specified requirements and monitors the progress of the supplier through periodic meetings and scheduled design reviews.

The activities in the R&M plan that address the user's R&M specifications, program matrix, and program planning worksheet, as discussed in Chapter Three, should be monitored by the user, preferably in concert with the machinery and equipment supplier. The plan outlines the specific activities to be undertaken to achieve improvements in reliability and maintainability. The supplier's progress toward accomplishment of those activities specified for the Design and Development phase will be targeted by the user for review.

Supplier's Role in the Design and Development Phase

During the Design and Development phase, the supplier performs various engineering studies and analyses to ensure that machinery and equipment designs meet the user's requirements. These may include tolerance studies, stress analyses, accelerated life testing, and incorporation of design margins into designs. Design reviews may be conducted internally and/or with the user to manage the evolutionary process of the design. The supplier may examine prior internal designs and historical data for Lessons Learned relative to desirable features to be retained, mistakes, and solutions. In this phase, the supplier develops suitable design verification and validation test plans that are agreed to by both the user and supplier to demonstrate compliance to the R&M and performance requirements. In addition to *designed-in reliability* and *designed-in maintainability*, the design should also provide for disassembly, transporting, and re-assembly in the user's plant. In consultation with the user, the supplier initiates the development of a list of recommended spare parts, based upon the reliability characteristics of components and inventory requirements, to be maintained by the user. As the design evolves, the impact of design changes on LCC objectives is assessed.

Recommended Practices in the Design and Development Phase

Table 5 lists the prevailing tools and techniques that are recommended practices for users and/or suppliers during the Design and Development phase. Each is discussed further in the text that follows Table 5.

Table 5. Recommended Practices in the Design and Development Phase

Tools and Techniques	User	Supplier
Design Margins and Stress Analyses		X
Machinery Components		X
Failure Mode and Effects Analysis		
Process FMEA	X	
Machinery FMEA		X
Fault Tree Analysis	X	X
Design Reviews		X
Tolerance Studies		X
Reliability Analysis and Predictions		X
Reliability Block Diagrams		X
Accelerated Life Testing		X
Maintainability Design Concepts		X
Maintenance Manuals and Preventive Maintenance Requirements		X
Spare Parts Lists & Spare Parts Inventory Plan	X	X
Accessibility	X	X
Diagnostics	X	X
Captive Hardware and Quick Attach/Detach	X	X
Spare Parts Management	X	X
Maintenance Procedures	X	X
Visual Management Techniques	X	X
Modularity		X

See Appendix E for information on other tools and techniques for reliability and maintainability.

Design Margins and Stress Analyses

All machinery suppliers should use some degree of conservatism in the design practices employed by the various engineering disciplines (e.g., electrical, mechanical, fluid power, etc.). For example, reliability is a function of applied stress to actual strength. When the applied stress exceeds the actual strength, failure occurs. Both stress and strength are statistically expressed as distributions, and therefore may not remain constant over time. Stress analyses done after design can provide evidence of overstress conditions. While this information is necessary, the recommended approach is to estimate the worst-case conditions applied to an evolving design and to provide a solution in advance in the form of an appropriate *design margin*. A *design margin* in electrical engineering is referred to as a "derating margin," and in mechanical engineering as a "safety factor."

Design Margins

For dependable operation, building derating margins and safety factors into designs should be done at the component level after taking into account the predominant stresses, including thermal stresses. Machinery designers should employ their own set of design margin criteria to ensure that the strength of a component exceeds the applied stress. Every supplier needs to evaluate the performance of its installed base of machinery and equipment and identify which components have high failure rates, and evaluate its historical design practices for effectiveness. Additionally, finite element analysis and similar electrical stress measurement methodologies can be used to assess the margins in designs and expected life. Incorporating adequate derating values and safety factors into designs is the responsibility of the supplier.

It should be noted that derating margins and safety factors must be assessed under worst-case conditions for both stress and strength because considering them only under nominal conditions can result in high failure rates. Strength data are available for many electrical and mechanical components and materials from their manufacturers. Stress is unique to each application and requires measurements on similar applications and calculations to account for differences of application.

Stress Analyses

For selected new designs and for existing designs that have been shown to be failure prone, it is important that the machinery supplier conduct stress analyses. Stress analyses use numerical analysis to determine the relationship between the strength of the component and the stress induced by the environment under worst-case conditions. Stress analyses will validate the effectiveness of the design margin criteria employed by the machinery supplier. The areas designated for stress analysis investigation should have been documented in the R&M plan. A number of engineering techniques can be used for stress analysis studies; finite element analysis, for example, is one such technique.

Machinery Components

Manufacturing machinery and equipment components should be selected based upon their ability to improve overall equipment reliability and maintainability. Specifications for and desired characteristics in reliability-sensitive or critical components should be defined by the machinery and equipment supplier and communicated to component suppliers. The R&M plan would identify how component suppliers and subcontractors are selected and, where applicable, the qualification process used to verify reliability, and how the integrity of the whole design is maintained including the critical components.

Failure Mode and Effects Analysis (FMEA) – Process and Machinery

Process FMEA

A Process FMEA is used to analyze the manufacturing and assembly processes. Potential product failure modes caused by the manufacturing and assembly process deficiencies are identified.

Users perform process FMEA on new manufacturing systems and may involve the manufacturing machinery and equipment supplier. Suppliers may perform a process FMEA for a turnkey program. Typically, this analysis would be performed early in the development of the new system with selected personnel from production operations and maintenance, and coordinated by manufacturing engineering. This activity focuses engineering and management attention on high-risk subsystems of the new system and requires upper management support during simultaneous engineering to ensure appropriate plant participation and that corrective action is taken. Suppliers also may perform a process FMEA on their internal processes that impact machinery and equipment reliability and maintainability.

Machinery FMEA (may be termed Tooling and Equipment FMEA or Plant Machinery and Equipment FMEA)

Machinery suppliers perform machinery FMEA on their machines. Similar to process FMEA, machinery FMEA focuses attention at the subsystem level for purposes of identifying potential failure modes and their impact on satisfactory performance of the machine. FMEA should be performed early in the design process to ensure that critical failure modes are eliminated from the design and that maintainability procedures are defined for the remaining non-critical failure modes. FMEA can be used for developing troubleshooting procedures. FMEA may be required by contract/purchase order. Refer to Appendix H for further information on FMEA.

Fault Tree Analysis (FTA)

FTA is a qualitative analytical model, which can be evaluated quantitatively if desired, that is conducted on an as-needed basis. FTA is a top-down method of systems analysis that is used to explore the possible occurrence of undesirable events (failures). An FTA is one of the few techniques that graphically depict the interaction of many factors and consider multiple fault conditions that, if occurring simultaneously, may lead to the undesirable event. FTA modeling starts with the undesirable event at the top and branches downward by logically evaluating the causal interrelationship of lower-level events or faults within the total system that could lead to the undesirable event. These interrelationships among faults, in combination or alone, determine the conditions that must be present to cause the undesirable event to occur. In contrast to FTA, which is used to analyze multiple undesirable events and multiple faults, FMEA is used to examine a single event. Safety aspects of a design can be analyzed using FTA early in the design phase to allow changes to be made quickly and easily. Fault trees are constructed graphically using common logic symbols to portray the undesirable events and faults.

Design Reviews

Design reviews are an integral part of the Design and Development phase. To be effective, multi-phased design reviews are desirable, consisting of at least a preliminary and a final design review. From an R&M standpoint, each review would focus on the R&M activities that are consistent with the evolutionary stage of the design. For example, during a preliminary design review, the focus should be on selecting design assurance activities, whereas during the final design review, the focus would be on the results of the R&M analyses. (See Chapter Three for additional information on design reviews.)

Tolerance Studies

Machinery suppliers should routinely conduct studies to ensure that electrical and/or mechanical tolerance stacking does not cause equipment failure or premature wear under the worst-case environmental conditions on the manufacturing floor.

Reliability Analysis and Predictions

A reliability prediction is a preliminary analysis of reliability that can be performed by generating a *reliability block diagram* at the component level. Using data collected on identical or similar equipment or using information from field data, gathered by the supplier or from other sources, component failure rate estimates can be incorporated into the reliability block diagram. From this information, machine reliability can be inferred through calculation. Machinery suppliers may perform these preliminary predictions of reliability to support bid responses and to substantiate that the specified reliability requirement and/or goal is attainable.

When reliability data from similar equipment are used, adjustments to the reliability Figures of Merit should be made to reflect any differences in use, complexity, and other operating characteristics. As additional component reliability data are accumulated, the reliability prediction should become more refined.

Reliability Block Diagrams

Reliability block diagrams are planning tools used to recognize how each equipment component contributes to the functional reliability of the equipment. The reliability block diagram used during the initial phases of design and development lays out the equipment design based on component interactions.

Reliability block diagrams are used to calculate system reliability based on component reliabilities. They can also be used to determine the minimum component reliability necessary to achieve desired system reliability.

The manufacturing of machinery and equipment is frequently a combination of serial and parallel components and processes. The analysis of complex combinations of serial and parallel elements uses formulas in combinations that mirror the equipment configuration to calculate system reliability.

Accelerated Life Testing

Accelerated life testing may be conducted on crucial components. Crucial components are those components having the greatest impact on the reliability of the machinery and equipment. Identification of crucial components should be made with input from the user as appropriate. Test conditions must be accelerated in an appropriate manner, using appropriate equations for accelerating the environment, so that the failure rate can be translated to normal use conditions. Results of accelerated life tests must be cautiously evaluated. There may be more than one failure mechanism, which result in different acceleration curves. Without careful attention to the accelerated methodologies, results can be understated or overstated.

Maintainability Design Concepts

The R&M plan should address maintainability design features that will minimize downtime for corrective and preventive maintenance. While designing equipment, the machinery designer should consider design features such as accessibility, modularity, standardization, reparability, testability (including built-in test equipment), and interchangeability. By understanding how the user will maintain the equipment, the designer can develop a *design for maintainability* that is in concert with the user's logistics and support system. Characteristics that should be addressed in designing for maintainability include:

- Repair with standard tools and test equipment
- Repair and maintenance within the skill levels of operating and maintenance personnel
- Accessibility with identifiable locations for predictive maintenance analysis (e.g., vibration sensors)
- Use of a maintainability assessment tool to compare different design alternatives. (See Appendix E for a sample of a maintainability assessment tool.)

Maintenance Manuals and Preventive Maintenance Requirements

The development of maintenance manuals, which include preventive maintenance requirements, is a necessary component of an effective R&M program. Maintenance manuals should consider the following:

- The skill levels of those who use the manuals in performing the maintenance functions
- Preventive maintenance requirements that are compatible with existing schedules and inventories, if known
- Estimates of time to carry out preventive maintenance actions
- Maintenance staffing recommendations, when machinery factors indicate unique requirements
- Type of training that may be required.

Spare Parts List and Spare Parts Inventory Plan

The manufacturing machinery and equipment supplier should make a recommended spare parts list available to the equipment user. Developing a list of recommended spare parts and preparing a spare parts inventory plan will help the user achieve increased machinery and equipment availability. Sourcing of spare or replacement parts, including consumable materials, should be managed by the user to ensure that the performance and capability of the manufacturing machinery and equipment are maintained at or above the supplier's specifications. Some things to consider include:

- Spare parts lists based on assembly or component part reliability characteristics
- Minimum spare parts inventory requirements
- Inventory locations and storage space requirements.

Accessibility

Accessibility means having sufficient working space around a component to diagnose, troubleshoot, and complete maintenance activities safely, effectively, and in a timely manner. Provision must be made for movement of necessary tools and equipment with consideration for human ergonomic limitations.

Diagnostics

Diagnostic devices that indicate the status of equipment should be built into manufacturing machinery to aid maintainability support processes. These devices can be as simple as a go/no-go visual display, or as sophisticated as a knowledge-based expert system with the capability of analyzing a problem and recommending the most likely solution. Diagnostic systems should indicate the specific component that needs replacement or repair.

It is desirable that diagnostic systems would have the capability of storing equipment performance data as permanent records for reliability analysis and supplier feedback. These systems can support the reliability growth management process. It is desirable for the output from diagnostic systems to be in a format that is compatible with commercially available database management software.

When component assemblies and subsystems are used to create a manufacturing system, hardware and software "hooks" are put in place in the Design and Development phase to facilitate integration of the diagnostics system in the build phase.

Captive Hardware and Quick Attach/Detach

Captive and quick attach/detach hardware provides for rapid and easy replacement of components, panels, brackets and chassis. The environment in which these devices are used may restrict the type of device used. Spare parts and replaceable subassemblies should also be configured with these devices pre-assembled. Examples include:

- Plate, anchor, and caged nuts
- Push and snap-in fasteners
- Clinch and self-clinching nuts
- Quarter-turn fasteners.

Spare Parts Management

Maintenance of manufacturing machinery and equipment requires a readily available supply of spare parts and supporting materials to operate, maintain, and service the equipment. Spare parts management identifies and analyzes the quantities of spare parts required to optimize inventory cost savings to the equipment user.

Plans for equipment support through spare parts management should begin with the supplier during the design process and continue with the user throughout the life cycle of the equipment. Consideration should be given to the lead time required to requisition, manufacture, and receive into inventory the required parts or materials to avoid the excessive costs of procuring replacement parts on an emergency basis.

Maintenance Procedures

Maintenance procedures describe in detail the adjustments, replacement, and repair of machine systems, subsystems, and component parts. The supplier will provide recommended preventive maintenance procedures at intervals based on time or machine cycle count. Maintenance requirements should be prioritized to enable the user to prioritize maintenance scheduling related to the criticality of the activity.

The maintenance procedures should be contained in service manuals or a computerized database reflecting the specific content and configuration of the equipment being supported. Exploded view illustrations, photographs, simplified assembly drawings, and parts lists relating to the required maintenance activities and procedures should be included wherever applicable. Technical information such as pressure settings, operation sequences, and moving part clearances should be included as appropriate.

Visual Management Techniques

Visual management techniques are used on machinery and equipment to bring the workplace awareness to a level that allows problems and abnormal conditions to be recognized quickly at a single glance. A visual management system enhances the equipment inspection process by allowing quick identification of safety, quality, environmental, equipment, and process abnormalities. A list of visual management techniques is in Appendix E.

Modularity

Modularity requires that designs be divided into physically and functionally distinct units to facilitate removal and replacement. Modularity mandates that components be designed as removable and replaceable units for an enhanced design with minimum downtime. Modular design concepts typically are thought of in terms of electrical black boxes, printed circuit boards, and other quick attach/detach electrical components. These concepts are also applicable to the mechanical elements of production equipment.

Modularity offers several advantages:

- New designs can be simplified and design time can be shortened by making use of standard, previously developed building blocks.
- Specialized technical skill will be reduced.
- Training of plant maintenance personnel is easier.
- Engineering changes can be made quickly with fewer side effects.

R&M Activity Matrix

Appendix C provides an example of a matrix of responsibilities for R&M program activities during the Design and Development phase. A user and/or supplier can use this matrix.

Chapter Five

R&M Activities During the Build and Install Phase

During the Build and Install phase, care must be taken to maintain inherent R&M, those R&M capabilities designed into the equipment. The supplier's manufacturing process variables affecting inherent R&M should be identified and targeted for control. Installation-related anomalies should be addressed prior to machine start-up to reduce the effect of compounding failures.

Introduction

During the manufacturing and assembly of the machinery and equipment, the achievement of R&M requirements that were incorporated into the design needs to be monitored. Issues that could affect R&M are communicated back to design engineering to ensure that any redesign includes R&M improvements. The supplier's manufacturing process variables affecting inherent R&M should be identified during this phase and targeted for control.

Events that occur during the Build and Install phase include:

- Training that starts and continues into the next phase (see Appendix I)
- The performance of acceptance testing, if required, prior to disassembly and installation at the user's facility
- Collection and documentation of R&M data. This event begins by documenting all issues encountered during this phase and during acceptance testing. The documentation is valuable for future reference and the data may provide candidates for continuous improvement.
- Identification of crucial assembly processes prior to disassembly, transporting, and reassembly of the machinery and equipment in the user's facility. The machinery and equipment has to be reassembled to the original build requirements.
- The documentation and elimination of any infant mortality (premature) failures that occur during the initial start-up. Every effort needs to be taken to eliminate failures during the installation and debugging period.

User's Role in the Build and Install Phase

During the Build and Install phase, the user should actively monitor the machinery and equipment construction process to help identify potential R&M-related manufacturing defects and participate in the development of maintenance procedures for the machinery and equipment. The user works with the supplier to ensure that the data collection requirements are met.

The user should actively participate with the supplier in the equipment run off to validate results and monitor the implementation of corrective actions to meet R&M requirements. This is also an excellent opportunity for users to begin data collection for future reference and training processes on the machinery and equipment.

The user may also perform a maintainability assessment during this phase to identify and eliminate any remaining maintainability issues.

Supplier's Role in the Build and Install Phase

During the Build and Install phase, the supplier manufactures and assembles the machinery and equipment to the design requirements established in the prior phase and performs appropriate testing to ensure that the equipment functions to user's specifications. Acceptance testing that may be required is performed and R&M and performance-related data are collected, logged, and analyzed for accuracy and required action. Additionally, maintenance procedures and schedules are finalized and documented in maintenance manuals, and care is taken to determine any special disassembly and re-assembly requirements. The supplier determines the training requirements and guides the user in identifying training needs and in gaining familiarity with the machinery and equipment prior to disassembly and shipment to the user's facility. Upon completion of installation at the user's facility, the machinery and equipment may be tested again to ensure that it has not been degraded during transport and installation.

As the Build and Install phase progresses, any changes made during the build and install process are evaluated for their impact on R&M objectives. The tools and techniques commonly used during this phase to ensure achievement of the R&M goals set for the new machinery and equipment are discussed below.

Recommended Practices in the Build and Install Phase

Table 6 lists the prevailing tools and techniques that are recommended practices for users and/or suppliers during the Build and Install phase. Each is discussed further in the text below.

Table 6. Recommended Practices in the Build and Install Phase

Tools and Techniques	User	Supplier
Dedicated Reliability Testing (Qualification Testing)		X
Preliminary Evaluation of Process Performance		X
Dry Run Testing		X
Reliability Data Collection During Acceptance Testing at the Supplier's Facility	X	X
Reliability Data Collection During Acceptance Testing at the User's Plant	X	X
Root Cause Analysis and Failure Analysis		X

See Appendix E for information on other tools and techniques for reliability and maintainability.

Dedicated Reliability Testing

If dedicated reliability testing was defined and detailed in the R&M plan, it is performed during this phase. Dedicated reliability testing is done for purposes of verifying attainment of specific R&M requirements. It should be noted that dedicated reliability testing could be costly and extensive in duration. A typical dedicated reliability testing program could extend to approximately four times the specified mean time between failure (MTBF) if that Figure of Merit is being assessed. Therefore, dedicated reliability testing should only be applied to crucial components. During the Concept and Proposal phase, the supplier and user mutually agreed upon the specified R&M requirements, specific reliability test objectives, and the duration of testing under normal operating conditions. When performing this type of testing, all failures should be documented and maintenance times recorded. When a contract calls for dedicated reliability testing to be performed on all units, it serves as acceptance testing; when it is to be done on only one unit, it may serve as qualification testing.

Preliminary Evaluation of Process Performance

The purpose of this preliminary testing is to aid in documenting component failure or process-related problems associated with infant mortality (premature) failure. This preliminary assessment of achievement of reliability and maintainability requirements is conducted in the supplier's facility before beginning acceptance testing of the machinery and equipment. All parts that are generated by a sample run are to be identified sequentially and may be required to be available for the user's review.

Dry Run Testing

Some type of performance testing, with or without parts, prior to acceptance testing, is an important measure of machinery readiness. Important machine parameters should be verified in advance of acceptance testing as operating within acceptable limits. A dry run for a set duration, twenty hours for example, without stoppage due to equipment failure, may serve as a demonstration of machinery readiness. Minor stoppages during the dry run may be allowed at the discretion of the user. All failures or stoppages, regardless of severity or consequences, should be documented and addressed.

Reliability Data Collection During Acceptance Testing at the Supplier's Facility

As a cost-effective alternative to dedicated reliability testing, reliability or failure data should be collected on all commercially purchased or specially manufactured components during acceptance testing at the machinery and equipment supplier's facility. These data provide an initial indication of reliability capability, and/or misapplication of components. Although testing time and failure event data may be limited, the information collected can be combined with process validation test results to develop a reliability benchmark. This benchmark then becomes the starting point for reliability growth through continuous improvement. In addition, the reliability of consumables (perishable tooling, coolants, lubricants, etc.) can be measured during the acceptance testing of associated machinery and equipment.

Reliability Data Collection During Acceptance Testing at the User's Plant

Reliability data should be collected during acceptance testing at the user's plant to verify that the known reliability characteristics have not been degraded during transport and installation. These data can be added to the database established during acceptance testing at the supplier's facility.

Root Cause Analysis and Failure Analysis

Suppliers are responsible for ensuring that root cause analyses of equipment failures will be performed by either themselves or their sub-suppliers. The results of these analyses will be reported back to the user so that the user and supplier can jointly determine how to best resolve any deficiencies. The requirement to perform root cause and/or failure analysis may be defined by contract, and the R&M plan defines the process by which failure issues are to be resolved during this phase.

R&M Activity Matrix

Appendix C provides an example of a matrix of responsibilities for R&M program activities during the Build and Install phase. The user and/or supplier can use such a matrix in their R&M planning.

Chapter Six
R&M Activities During the Operation and Support Phase

In the Operation and Support phase, the user is expected to implement a system of R&M data collection, analysis, and feedback. The supplier uses feedback to improve the R&M of existing as well as new machinery and equipment designs.

Introduction

The machinery and equipment that was delivered, installed, and tested is now producing product and is to continue to produce product throughout its operational life. Data collection, feedback, and preventive maintenance activities are the significant activities in this phase, in that each is meant to support the continuous operation and improvement of the machinery and equipment.

Data collection mechanisms should be in place and operational as agreed upon by the user and sup plier. Data collected during this phase may be used to validate R&M requirements. Performance and R&M information collected during this phase should lead to R&M growth and continuous improvement for the next-generation design. Additionally, continuous improvement and maintenance programs may be augmented by integration of R&M improvements in existing designs because of failure data analyses and the corrective actions taken during this phase.

During this phase, planned maintenance should be performed as agreed. If the manufacturing machinery and equipment supplier is to optimize its R&M initiative, the supplier and its sub-suppliers need to have access to user maintenance records and R&M databases. The sharing and analysis of user maintenance information is an effective communication activity that can yield long-term benefits to both user and supplier in their continuous improvement activities, and in the supplier's and its sub-suppliers' next-generation equipment design activities.

User's Role in the Operation and Support Phase

In this phase, the user operates the machinery and equipment and performs maintenance activities as planned and as defined in supplier maintenance manuals. R&M and performance information is collected and acted upon, involving at times supplier personnel. To ensure continuous improvement of machinery and equipment designs, R&M data as well as user corrective actions and failure report summaries are provided to suppliers for analysis and consideration in future product designs.

Supplier's Role in the Operation and Support Phase

Using the reliability and maintainability and machinery and equipment performance information provided by users, suppliers should develop a means for monitoring, analyzing, and incorporating improvements into current and future designs to reduce or eliminate R&M deficiencies and provide reliability and maintainability growth.

Recommended Practices in the Operation and Support Phase

Table 7 lists the prevailing tools and techniques that are recommended practices for users and/or suppliers during the Operation and Support phase. Each is discussed further in the text below.

Table 7. Recommended Practices in the Operation and Support Phase

Tools and Techniques	User	Supplier
Data Collection, Analysis and Feedback	X	X
Planned Maintenance	X	
Reliability and Maintainability Growth	X	X
Failure Reporting, Analysis, and Corrective Action System (FRACAS)	X	X
Data Exchange	X	X
Suggested Data Feedback Model	X	

See Appendix E for information on other tools and techniques for reliability and maintainability.

Data Collection, Analysis, and Feedback

In this phase, the user implements a system of R&M data collection, analysis, and feedback that was designed and installed during the earlier phases. The efficient and continuous collection and feedback of equipment operation data are crucial to a successful R&M program. Accurate and consistent recording of failures, symptoms, and corrective actions is far more important than the media used for data collection (paper or electronic). A paper system is still an effective means of recording meaningful and accurate data.

Data needed to report failures and to calculate Figures of Merit are generated by a user's operation and maintenance activities. This information is then used to support continuous improvement. Users and suppliers should work together to properly analyze failures and develop corrective actions.

Planned Maintenance

For equipment to meet its R&M objectives consistently, it must be properly maintained. To accomplish this, the user should have an effective planned maintenance program in place that may include:

- Participation by those responsible for operating the machinery and equipment
- Preventive maintenance
- Predictive technologies
- A proactive organization trained in R&M practices.

Reliability and Maintainability Growth

When required by contract, suppliers may be expected to participate in the user's reliability and maintainability growth (continuous improvement) process during user manufacturing operations by assisting in the root cause analysis of the critical failures that occur. Through the identification and communication of both failure causes and implemented corrective actions, the inherent reliability and maintainability of equipment can be improved. Reliability and maintainability growth can be effective in reducing warranty costs and improving future products. It is strongly recommended that suppliers participate in a reliability and maintainability growth process for future designs even when it is not required by contract or purchase order.

Failure Reporting, Analysis, and Corrective Action System (FRACAS)

A Failure Reporting, Analysis, and Corrective Action System (FRACAS) provides for the orderly recording, transmission, and use of failure data. It furnishes: failure reporting to establish a historical database, failure analyses that may be used to determine the causes of failure, and documentation of corrective actions, which may be used to decrease the incidence of particular failure causes. A failure report database allows for the identification of patterns of failures and rapid resolution of problems through rigorous failure analysis. Resolution of problems (corrective actions) promotes the reliability growth of equipment in the field as well as improved reliability for new equipment. In addition, by incorporating maintainability data into the database, repair performance can be monitored.

Under a FRACAS system, the supplier should develop a means of summarizing and analyzing failure data obtained at both supplier and user facilities as a means of promoting R&M improvement activities. Reliability growth is accomplished by using failure data from this phase to improve new machine designs in the next Design and Development phase in the continuum of life cycle phases.

Data Exchange

The supplier and user jointly developed a system for transferring user machinery and equipment performance and maintenance data in the earlier life cycle phases and are now intent upon the continuous and effective exchange of that data throughout the operational life of the machinery and equipment. The data being exchanged during this phase should contain sufficient information to allow calculation of the R&M Figures of Merit. The format of the data should be as mutually agreed upon and provide sufficient detail so that the supplier can isolate a problem and develop corrective action.

This exchange of data benefits both suppliers and users. The benefits to the machinery and equipment supplier includes a clear understanding of the performance of their machines in user facilities, insight into the reduction of warranty costs, and improved user satisfaction. The benefits to the machinery and equipment user are improved machinery performance and a more responsive supplier. As data exchange technologies change over time, both user and supplier need to remain open to considering new collection and transmission technologies, and flexible in their exchange relationship to adopt those technologies that are both economical and beneficial to the mutual use of the data. See also Appendix G.

Suggested Data Feedback Model

Figure 7 shows a sample flow chart that tracks documentation of a failure within a typical manufacturing plant and to the supplier. This feedback model facilitates user communication to suppliers and this chart depicts the supplier's role in the data feedback system. See also Appendix G for additional information on data tracking and feedback systems.

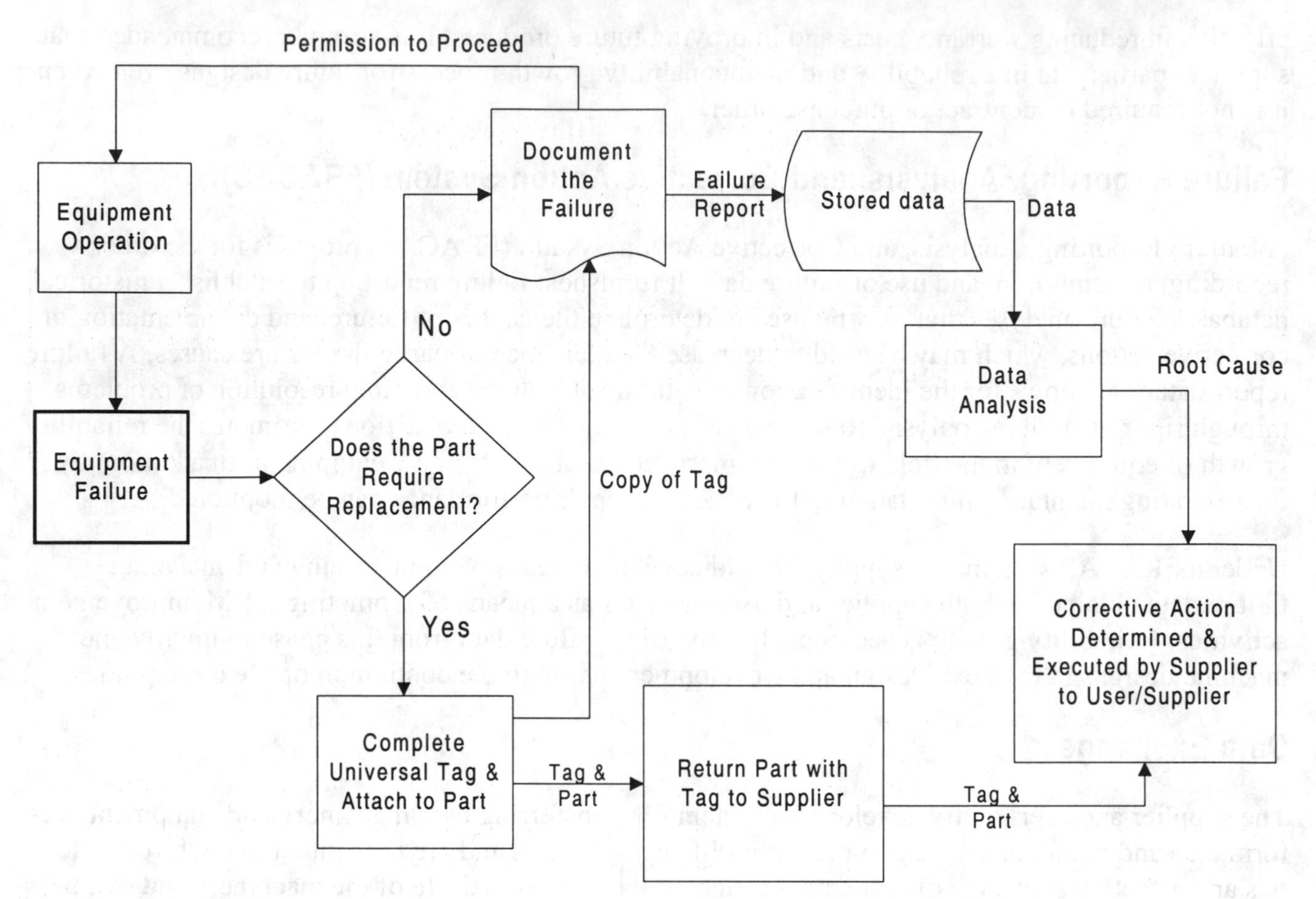

Figure 7. Basic Failure Documentation System

R&M Activity Matrix

Appendix C provides an example of a matrix of responsibilities for R&M program activities during the Operation and Support phase. The user and/or supplier can use such a matrix in their R&M planning.

Chapter Seven
R&M Activities During the Conversion or Decommission Phase

The Conversion or Decommission phase marks the end of the expected life of a piece of machinery or equipment. When either conversion or decommission occurs, feedback from the user plant should be considered for R&M growth and continuous improvement in future generations of machinery.

Introduction

During this phase, the machinery and equipment may be converted or decommissioned. When converted, machinery and equipment may be reworked, retrofitted, rebuilt, or remanufactured. When decommissioned, machinery and equipment may be removed and sold, scrapped, stored for future use, or transferred and reinstalled in a different location. R&M activities occurring in the conversion or decommission phase are dependent on the disposition of the machinery and equipment and its age. For example, decommission with the intention to reuse a machine that is of recent manufacture (less than 10 years old) may have data collection and analysis performed, while decommission of an aged machine (greater than 15 years old) that is being scrapped may not require any data collection at all.

When equipment is being evaluated for conversion, recent R&M data can provide insight into design improvements beneficial to its intended use. When the conversion of machinery and equipment incorporates a new or modified design, performing R&M activities such as those discussed in the Concept and Proposal phase and Design and Development phase can be appropriate (see Chapters Three and Four). Any existing FMEA can be reviewed and updates can be performed as deemed necessary to provide input to the conversion activity. The R&M techniques discussed in previous chapters can be applied during various conversion processes (chapter cross-references for various types of conversions are given later in this chapter).

Decisions to convert, decommission, or improve manufacturing machinery and equipment should involve all disciplines at the plant, to ensure that the appropriate information gathering and analysis is completed prior to disposition of the manufacturing machinery and equipment.

User's Role in the Conversion or Decommission Phase

When machinery and equipment are slated for conversion or decommission, machinery and equipment users can assemble previously collected data or, minimally, can prepare an anecdotal performance review of the machinery and equipment. Previously collected data that is retained by a user will depend upon company policies. These data can include failure summaries, maintenance logs, Lessons Learned summaries, etc. The effort to assemble and revisit previously collected data or to prepare a post-performance anecdotal review is to facilitate understanding the past performance of the machinery and equipment for input into requirements for conversion and for new-generation equipment designs. If previously collected data are not retained or are not available for use in conversion activities or new designs, a supplier's ability to perform some R&M activities, which otherwise might contribute to the optimization of the conversion or new design, will be diminished. For conversion projects, a review of process FMEA and machinery FMEA for improvement opportunities is desirable. In addition, a review of preventive maintenance practices including the effectiveness of any predictive and detective maintenance procedures used for maintaining or improving machinery uptime can provide important input for the design of new equipment. If this information can be made available, suppliers can refine their preventive maintenance recommendations and develop designs better suited to implementation of effective preventive maintenance programs.

Supplier's Role in the Conversion or Decommission Phase

Suppliers should seek to obtain user data reports, or if such reports are unavailable, a post-production anecdotal review for use in improving future machinery and equipment designs and improving performance of converted equipment. For conversion projects of machinery and equipment, a review of process and machinery FMEA for improvement opportunities should be conducted whenever these data are available. Additionally, if a conversion is contracted to the original machinery and equipment supplier, supplier-generated data may be collected and reviewed. The additional data that can be brought to the improvement process by the original machinery and equipment supplier include design FMEA, build data, warranty data, field service data, etc.

Recommended Practices From Earlier Life Cycle Phases

Table 8 cross-references the guideline chapters containing recommended practices relevant to machinery conversions. R&M planning for machinery conversions is similar to the planning for new machinery and equipment depending on the condition of the machinery and its intended application.

Table 8. Applicable Guideline Chapters for Machinery Conversions

Type of Conversion	New or Modified Design Incorporated	Applicable Guideline			
		Chapter Three Concept and Proposal	*Chapter Four* Design and Development	*Chapter Five* Build and Install	*Chapter Six* Operation and Support
Retool	Yes	X	X	X	X
	No			X	X
Remanufacture	Yes	X	X	X	X
	No			X	X
Rebuild	Yes	X	X	X	X
	No			X	X
Retrofit	Yes	X	X	X	X
	No			X	X
Rework	Yes	X	X	X	X
	No			X	X
Clean and Scrape	N/A			X	X

Appendix A
Life Cycle Phases and Life Cycle Costs

Life Cycle Cost

Life Cycle Cost (LCC) refers to the total cost of a system during its life cycle. LCC is the sum of the costs incurred in all phases of the life of a machine. Figure A-1 illustrates five life cycle phases for machinery and equipment. The costs incurred during the first three phases, Concept and Proposal, Design and Development, and Build and Install, and the final phase, Conversion or Decommission, are considered non-recurring costs. Costs incurred during the Operation and Support phase are considered recurring costs, as they are incurred throughout the operational life of the machinery and equipment. The Operation and Support phase typically contributes the majority of costs represented in LCC.[1] The cost of operation and support of manufacturing machinery and equipment over the life cycle is usually much more than the initial acquisition cost.

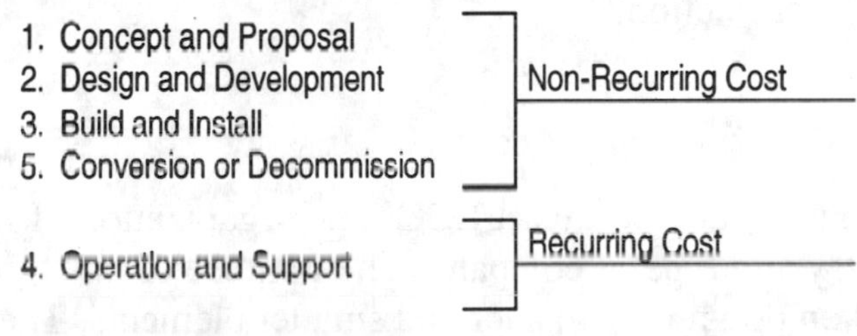

Figure A-1. Life Cycle Phases for Machinery and Equipment

An LCC analysis is an economic evaluation used to explore the:

- Economic feasibility of various production machinery and equipment alternatives for a specific application prior to acquisition; or the
- Economic feasibility of changing key cost factors that affect a system's LCC; or the
- Accumulated cost of a system at some point during its life cycle.

Applying the LCC analysis concept simply means identifying and summing all costs associated with a system's life cycle based on a set of assumptions about that life cycle. When LCC is calculated prior to acquisition, costs may be projected based on historical costs of similar systems, and when calculated for present value, costs should be summed based on actual cost experience. Supplier inputs to define base data, such as estimated numbers of spare parts, time to repair, etc., may be obtained, but users will want to control the LCC methodology and the cost factors applied to ensure that the LCC outcome depicts an accurate expectation for the user.

---

[1] For example, the Department of Defense reported that its operation and support costs for some of its systems was 72% of the total life cycle cost incurred during its service lifetime. Source: Principal Assistant Deputy Under Secretary of Defense for Logistics, Mr. Roy Willis, 1996.

Application of a Life Cycle Cost Analysis

Application of an LCC analysis requires the following steps:

1. Analyze and define the total life cycle phases for the system.
2. Define the expected life of the machinery and equipment based on a reasonable operational life period.
3. Determine the cost factors and relationships for each of the phases. Appropriate discount factors may be applied for present value analysis.
4. Consider the time related factors for costs such as inflation and rate of return, and hidden cost factors such as cost of money, cost of lost production, etc. if applicable.
5. Calculate the LCC by formulating a mathematical model.
6. Analyze the calculation to determine if the LCC calculation is particularly influenced or skewed by any single cost factor or cluster of like cost factors (e.g., inordinately high utility costs or inordinately low plant-related capital costs) and take these factors into consideration in the final interpretation of the LCC projection.

LCC Summary

Figure A-2 illustrates various costs that impact LCC. The organization of these cost elements may change based on the industry and type of company. The purpose of this figure is to show that cost elements are typically broken down into smaller and smaller elements. These can be tracked individually and incorporated into a mathematical model and then computed to form larger segments of cost, which in turn can be summed to obtain LCC. Brief descriptions of these cost elements follow.

Cost Elements That May Contribute to LCC

Acquisition Costs

Descriptions of the types of cost that contribute to acquisition costs are included below.

- *Purchase price:* The delivered price of the manufacturing machinery and equipment excluding transportation costs. This purchase price should account for currency differences and cost of money associated with payment schedules.
- *Administration and engineering costs:* Personnel, travel and runoff costs.
- *Installation and launch costs:* The costs uniquely associated with the installation of the manufacturing machinery and equipment, such as unique plant alternations, connection and debugging, etc.
- *Training costs:* The costs associated with training the work force to operate or maintain the particular machinery and equipment.
- *Conversion costs:* The direct and indirect costs associated with a planned conversion of the machinery and equipment from manufacturing one product to another product during its expected life.

- *Transportation costs:* The costs associated with the movement of machinery and equipment from its point of manufacture to user location.

Operating Costs

A further description of the types of cost that contribute to operating costs is included below.

- *Direct labor costs (operating):* The total cost of direct labor to operate the manufacturing machinery and equipment over its expected life.
- *Utility costs:* The total consumption cost of utilities over the expected life of the manufacturing machinery and equipment. Includes utilities such as air, steam, electricity, gas, and water.
- *Consumable costs:* The costs of all consumable items used by the machinery and equipment over the expected life. Cost items to consider are coolant, lubricants, filter media, etc.
- *Waste-handling costs:* The costs of collecting and disposing of waste products associated with the operation of the equipment.
- *Cost of spare parts inventory:* The cost of carrying and inventorying spare parts to support the machinery and equipment.

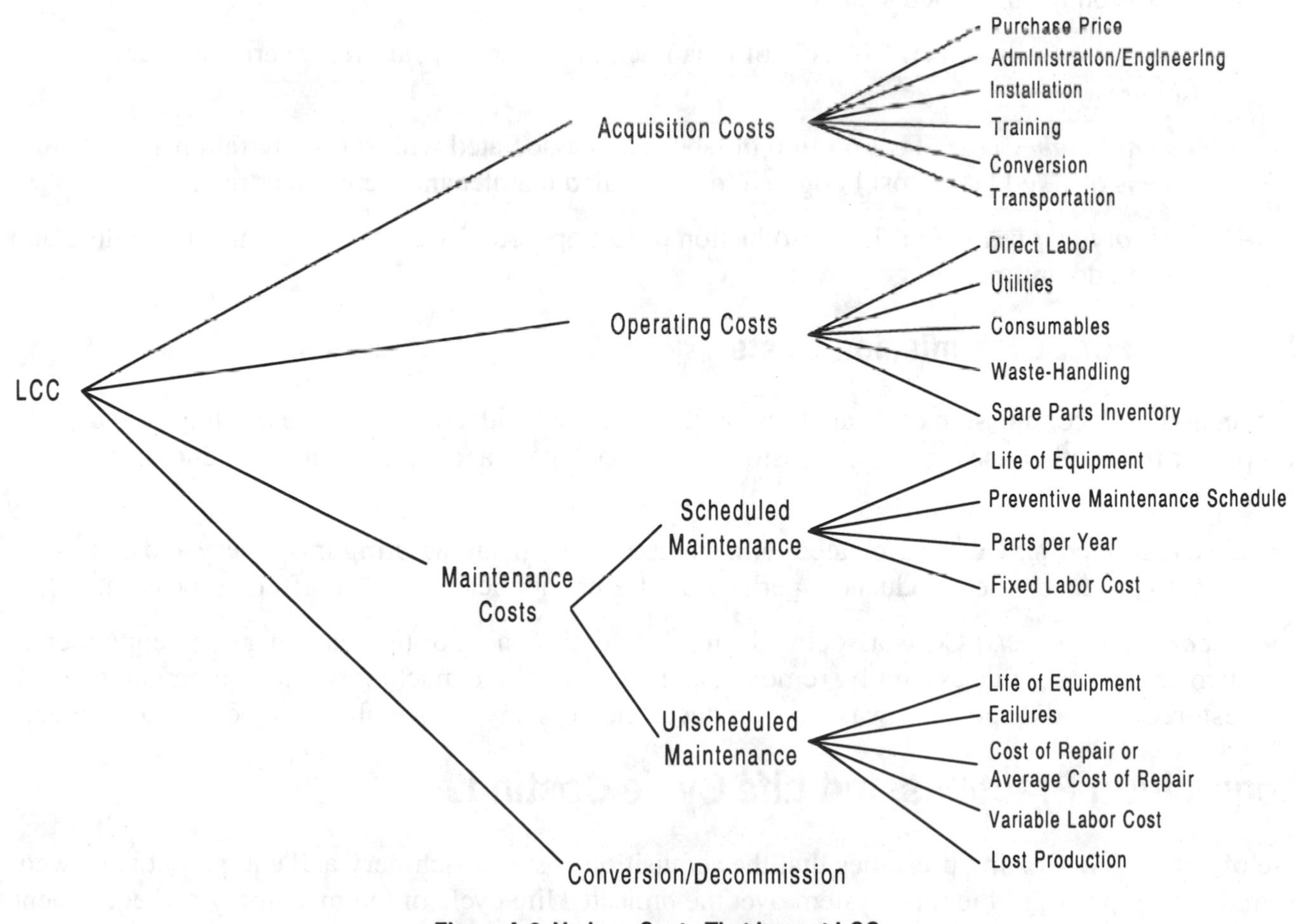

Figure A-2. Various Costs That Impact LCC

Maintenance Costs

Maintenance costs include costs associated with scheduled and unscheduled maintenance as discussed below.

- *Scheduled maintenance costs:* Costs of material and labor associated with the preventive maintenance schedule during the expected life of the machinery and equipment.
  - *Life of equipment:* The length of time the equipment is expected to produce until it is converted or decommissioned.
  - *Preventive maintenance schedule:* The schedule of recurring maintenance actions to ensure continued operation of the equipment. The machinery and equipment supplier prescribes the maintenance actions and frequency of occurrence.
  - *Parts per year:* The cost of parts consumed in this activity in one year.
  - *Fixed labor cost:* Cost associated with maintaining a pool of skilled labor to service scheduled maintenance.
- *Unscheduled maintenance costs:* Cost of material and labor associated with failures during the expected life.
  - *Cost of repair:* Cost of labor and parts necessary to restore the equipment to operable condition from a failed state, or
  - *Average cost of repair:* The cost of repair actions caused by failures, averaged over several failures.
  - *Variable labor cost:* That portion of labor costs associated with servicing failures that is in excess of fixed labor cost budgeted for scheduled maintenance, e.g., overtime.
  - *Cost of lost production:* Lost production or lost opportunity costs due to machine failure and related downtime costs.

Conversion or Decommission Costs

Conversion or decommission costs are those costs associated with converting the machinery and equipment for another use, or decommissioning the machinery and equipment at the end of its useful life.

- *Conversion costs:* Costs associated with converting the manufacturing machinery and equipment to produce other products in perhaps a different application and a different location.
- *Decommission costs:* Costs associated with decommissioning of the machinery and equipment, such as the cost of disassembly, removal, and disposal of the machinery and equipment, site restoration, and disposal of waste by-products, etc. less any recoverable scrap or salvage value.

Acquisition Practices and Life Cycle Costing

Use of Life Cycle Costing presumes that the acquisition cost of machinery and equipment is viewed from the perspective of the total system over the projected life cycle of the machinery and equipment. Use of Life Cycle Costing allows for consideration of additional factors, which add to system cost, in the acquisition decision. The practice of considering only initial purchase price in the purchasing decision can be modified to encompass an evaluation and comparison of the total costs to be incurred during the life cycles for various systems under consideration. Additionally, Life Cycle Costing allows for 'what if' evaluations when considering specification changes to a planned acquisition. See sensitivity analyses below.

Sensitivity Analyses

A sensitivity analysis is an analytical technique based on monitoring how outputs vary as inputs are systematically varied. Once an LCC analysis methodology has been established, it can be reapplied using varying inputs to test accuracy of the model, accuracy of select input data, and the impact of specific variables on LCC. By introducing variances in selected input data, one can determine what cost element(s) have the greatest influence on the LCC outcome. This type of analysis can be used to evaluate alternatives ('what if') and risks associated with various alternatives. Such analyses can assist a user in setting specifications for a planned acquisition, in selecting plant sites and facilities for system installations, in determining staffing requirements and mixes, and in evaluating present value against earlier LCC analyses, among others.

Sensitivity Analyses

Appendix B
Tailored R&M Program Matrices

The following three examples (Tables B-1 throught B-3) depict how R&M quantitative and qualitative requirements may be included in a machinery and equipment specification. It is the responsibility of the user, with help from the supplier community, to ensure that its specification documents are properly tailored to be consistent with user product requirements to avoid placing too little or too much emphasis on R&M. Proper tailoring of these requirements by the user ensures a cost-effective application of R&M program elements.

Table B-1. R&M Program Matrix for Existing Design—Single Machine Operation

Equipment Name:__________	Concept and Proposal Phase	Design and Development Phase	Build and Install Phase	Operation and Support Phase	Conversion or Decommission Phase
Reliability & Maintainability Requirements	X				
Machinery Use	X				
Failure Definition	X				
Duty Cycle	X				
Machinery Environment	X				
Continuous Improvement Information	X				
Life in Terms of Throughput	X				
Data Collection System	X				
Design Review					
Stress Analysis & Design Margins					
Maintainability Design					
Preventive Maintenance Requirements		X			
Machinery Components					
Reliability Predictions*		X			
Accelerated Life Tests					
Failure Modes and Effects Analysis*		X			
Dry Run Testing					
Dedicated Reliability Testing (Qualification Testing)					
Acceptance Testing at Supplier Facility					
Acceptance Testing at User Plant					
Reliability / Maintainability Growth			X		
FRACAS			X		
Data Feedback			X		X

* If not previously performed, corrective action should be taken for high-failure-rate items and must be taken for safety-related items.

SAMPLE

Table B-2. R&M Program Matrix for Existing Design With Major Modifications

Equipment Name:____________	Concept and Proposal Phase	Design and Development Phase	Build and Install Phase	Operation and Support Phase	Conversion or Decommission Phase
Reliability & Maintainability Requirements	X				
Machinery Use	X				
Duty Cycle	X				
Failure Definition	X				
Machinery Environment	X				
Continuous Improvement Information	X				
Life in Terms of Throughput	X				
Data Collection System	X				
Design Review*	X	X			
Stress Analysis & Design Margins		X			
Maintainability Design		X			
Preventive Maintenance Requirements		X			
Machinery Components		X			
Reliability Predictions**		X			
Accelerated Life Tests					
Failure Modes and Effects Analysis**		X			
Dry Run Testing			X		
Dedicated Reliability Testing (Qualification Testing)					
Acceptance Testing at Supplier Facility			X		
Acceptance Testing at User Plant			X		
Reliability / Maintainability Growth				X	
FRACAS				X	
Data Feedback				X	X

* Modified design only.

** Modified design and existing design if not previously performed.

SAMPLE

Table B-3. R&M Program Matrix for New Concept and Design (All Complexities — Single Machine to Transfer Line)

Equipment Name:__________	Concept and Proposal Phase	Design and Development Phase	Build and Install Phase	Operation and Support Phase	Conversion or Decommission Phase
Reliability & Maintainability Requirements	X				
Machinery Use	X				
Duty Cycle	X				
Failure Definition	X				
Machinery Environment	X				
Continuous Improvement Information	X				
Life in Terms of Throughput	X				
Data Collection System	X				
Design Review	X	X			
Stress Analysis & Design Margins		X			
Maintainability Design		X			
Preventive Maintenance Requirements		X			
Machinery Components		X			
Reliability Predictions		X			
Accelerated Life Tests		X			
Failure Modes and Effects Analysis		X			
Dry Run Testing			X		
Dedicated Reliability Testing (Qualification Testing)*			X		
Acceptance Testing at Supplier Facility			X		
Acceptance Testing at User Plant			X		
R&M Growth				X	
FRACAS				X	
Data Feedback				X	X

* Critical components and/or assemblies.

SAMPLE

Appendix C

Sample R&M Activity Matrix

This matrix lists the various R&M activities occurring in each life cycle phase and identifies who, the user or supplier, is typically responsible for leading or supporting the activity and the organizational function commonly assigned that responsibility.

Key to Matrix: L = Lead Responsibility S = Support the Process I = Input to the process A = Approval Responsibility

Life Cycle Phases and Activities	Supplier	Users			
		Manufacturing Engineering	Plant Operations (Workgroups & Engineering)	Purchasing	Other ______
Phase I – Concept and Proposal					
Plan R&M Requirements		L	S	S	
Assess Lessons Learned	S	L	S		
Specify Reliability & Maintainability Requirements		L	S	S	
Define Machinery Use	S	L	S	S	
Define Duty Cycle		L	S	S	
Define Machinery Environment		L	S	S	
Share Continuous Improvement Information with Suppliers		L	S		
Define Life in Terms of Throughput		L	S	S	
Specify Data Collection System	S	L	S	S	
Prepare Procurement Documents		S	S	L	
Prepare LCC Objectives	I	S	L		
Prepare & Submit R&M Plan	L	A	S	S	
Conduct Design Reviews	L	A	S	S	
Prepare LCC projections	I	L	S		
Phase II – Design and Development					
Apply Stress Analysis & Design Margins	L	S			
Select Machinery Components	L	S	S		
Create Maintenance Manuals	L	S	S		
Use Failure Modes and Effects Analysis	L	S	S		
Use Fault Tree Analysis	L	S			
Conduct Design Reviews	L	S	S		
Conduct Tolerance Studies	L	S			
Make Reliability Predictions	L	S			
Use Reliability Block Diagrams	L	S			
Perform Accelerated Life Tests	L	S			
Apply Maintainability Design Concepts	L	S	S		
Conduct Maintainability Assessment	S	S	L		
Define Preventive Maintenance Requirements	L	S	S		
Define Maintenance Procedures	L	S	S		

Key to Matrix: L = Lead Responsibility S = Support the Process I = Input to the process A = Approval Responsibility

Life Cycle Phases and Activities	Supplier	Users			
		Manufacturing Engineering	Plant Operations (Workgroups & Engineering)	Purchasing	Other ______
Prepare Maintenance Manuals	L	S	S		
Create Spare Parts List	L	S	S		
Create Spare Parts Inventory Plan	L	S	S		
Create Spare Parts Management Plan	S	S	L		
Define Accessibility Parameters	L	S	S		
Define Diagnostic Systems	L	S	S		
Consider Captive Hardware and Quick Attach/Detach Hardware	L	S	S		
Consider Visual Management Techniques in Machinery Design	L	I	I		
Consider Modularity in Machinery Design	L	I	I		
Consider LCC Impact in Machinery Design	L	I	I		
Phase III – Build and Install					
Conduct Dedicated Reliability Testing (Qualification Testing)	L	A	A		
Perform Preliminary Evaluation of Process Performance	L	I	I		
Dry Run Testing	L	A	A		
Perform Acceptance Tests	L	A	A		
Collect Reliability Data During Acceptance Testing at Supplier Facility	L	S	S		
Collect Reliability Data During Acceptance Testing at User Plant	S	S	L		
Conduct Root Cause / Failure Analysis	L	S	S		
Collect LCC Data	I	L	S		
Phase IV – Operation and Support					
Collect Data, Analyze Data, and Provide Feedback	S	S	L		
Conduct Maintenance as Planned	S	S	L		
Conduct R&M Growth Program	S	S	L		
Use FRACAS	L	S	S		
Exchange Data	L	S	S		
Collect LCC Data	I	S	L		
Phase V – Conversion and/or Decommission					
Characterize Equipment R&M	S	L	S		
Collect All Data and Lessons Learned	S	S	L		
Total LCC and Compare to LCC Objectives	S	S	L		
Adjust LCC Methodology if required	I	L	S		

Appendix D
R&M Program Planning Worksheet

(to be completed by machinery and equipment user)

Date: ______________________

1. Facility/Machine Name: ______________________
2. Process/Operation Number: ______________________
3. Brass Tag Number: ______________________
4. Supplier: ______________________
5. Model Year: ______________________

   Plant: ______________________

   Product: ______________________

Lessons Learned

6. Identify R&M Lessons Learned on same or similar machinery and equipment now in operation.

____

____

____

7. List the most important environmental concerns affecting R&M and machine performance:

Environmental	Range	Rate of Change

8. Identify process-related R&M concerns for this application, from similar applications:

____

____

____

9. List the R&M target values for this machine/facility (select appropriate Figures of Merit):

Factor	Target Value

Action Plan

10. For each equipment life cycle phase, list the R&M tools and techniques to be used on this program:

Concept and Proposal

Specific Action/Tool/Technique To Be Applied	Responsible Personnel

Design and Development

Specific Action/Tool/Technique To Be Applied	Responsible Personnel

Build and Install

Specific Action/Tool/Technique To Be Applied	Responsible Personnel

Operation and Support

Specific Action/Tool/Technique To Be Applied	Responsible Personnel

Appendix E
R&M Tools and Techniques

Part I: Reliability Tools and Techniques

This appendix provides information about selected reliability tools and techniques in addition to those referenced within the guideline. This discussion provides an introduction to these reliability tools and techniques and is not meant to provide the level of detail that a reliability engineer would need.

Computer Simulation

Computer simulations are descriptive modeling techniques used to study manufacturing systems to confirm system configuration and performance expectations. The simulation process should begin prior to the construction of the manufacturing machinery and equipment in the Concept and Proposal and the Design and Development phases, to enable the findings of the process to be implemented in the system design. The need for simulation varies depending on the complexity and expectations of the system. Simulation programs can provide valuable visual feedback to both the manufacturing machinery and equipment supplier and user. Such programs can be a valuable communication tool, translating concepts into visualizations. A cross-functional team should be used to support the implementation of a computer simulation program.

From an engineering perspective, the objective of a simulation analysis is to determine the predicted performance of the total manufacturing system in response to bottlenecks, breakdowns, and resource constraints within the system. By optimizing the process flow, selecting and sizing automation buffers, and specifying support resources in accordance with the recommendations from a cross functional team, system efficiencies can be improved and costs reduced. With accurate data and careful configuration of the model, the system can be visualized "in production" before it is built.

Potential areas for optimization that lend themselves to computer simulation include:

- Throughput of components
- System efficiency
- Location of queues within the system
- Location of production bottlenecks
- Inspection frequencies
- Tool change management recommendations
- Batch sizes for multiple part systems
- Maintenance management program recommendations
- System staffing requirements
- Staffing requirements by trade classification:
  - — Operators
  - — Maintenance personnel
  - — Quality inspectors.

Computer simulation support data must be carefully chosen and applied. The simulation can support and document the manufacturing system, and should be updated as often as required to accurately depict the actual system configuration. Equipment change recommendations from the reliability and maintainability growth management process can be simulated at a minimal cost to determine the net effect of the changes before making the final commitment of resources (personnel and capital) to implement the changes.

Environmental Considerations

When considering the design criteria for any process or piece of equipment, the manufacturing machinery and equipment supplier needs to make sure that the R&M planning process addresses the plant environment. The environment conditions that the process or equipment will be expected to encounter must be thoroughly documented, including not only the levels but also the rates of change. Some environmental factors are:

- Temperature
- Mechanical shock
- Immersion or splash
- Electrical noise
- Electromagnetic fields
- Ultraviolet radiation
- Humidity
- Corrosive materials
- Pressure or vacuum
- Contamination and their sources
- Vibration
- Utility services.

In addition to accounting for environmental conditions that could affect the process or equipment, consideration must be given to the effects that the process or equipment will have on the plant, the work place, and the earth's environment. The design of any process or piece of equipment has to consider Environmental Protection Agency (EPA) issues and Occupational Safety and Health Administration (OSHA) regulations.

Ergonomic Considerations

Ergonomics is an applied science that deals with the interface between people and their work place. Ergonomics considers the characteristics and capabilities of people in the design and arrangement of the work space, tools and equipment, work methods, and facilities, so that people and things can interact most effectively and safely. Applying ergonomics to the design considerations for manufacturing machinery and equipment can impact:

- Maintainability
- Operability
- Ease of inspection (to detect abnormalities)
- Access points.

Structured Problem Solving

Structured Problem Solving is a logical, systematic approach to resolving problems. Most large corporations have adopted a formal methodology for structured problem solving. Among the more prominent are DaimlerChrysler's 7D, Ford's 8D, General Motor's IAPIE, Cummins 7 step, and Crosby's 5-step approach to problem solving. The problems addressed may involve hardware failures, operational or process problems, maintenance difficulties, etc. or some combination of these problems.

Purpose

The purpose of structured problem solving is to provide a framework to resolve problems with manufacturing machinery, equipment, and tooling in a systematic, orderly manner.

Implementation

Structured problem solving methodologies have the following in common: problem identification, solution planning, problem containment (quick fix), identification of the root cause of the problem, developing and implementing a corrective action (removal or reduction of the frequency of occurrence of the root cause), follow up, and evaluation of the effectiveness of the corrective action. Performing structured problem solving as a team effort including users, suppliers, and sub-suppliers can have expanded benefits over the same methods with limited participation, especially for large problems having multiple potential root causes.

- *Problem Identification:* before any problem can be solved, the nature of the problem should be described in writing.
  - — Describe the data.
  - — Specify the nonconformance.
  - — Determine the size of the problem in numbers or cost.
- *Solution Planning:* ensures an orderly approach to the removal of the root causes of the problem.
  - — Identify the necessary resources: personnel, material, funding.
  - — Determine a measure of completion that ensures the problem has been solved.
  - — Establish a completion target date that will help to ensure steady progress is being made toward resolution of the problem.
- *Problem Containment:* temporary measures implemented to keep operating, taken before the formal problem solving is started.
  - — Document as part of the process so that an accurate picture of how the problem was solved is recorded.
  - — Recognize that this effort is not the permanent solution.

- *Root Cause Identification:* an analytical process used to identify the basic reasons (causes, mechanisms, etc.) for a problem, failure, non-conformance, process error, etc.
  - — Recognize that multiple root causes may exist, especially for complex problems.
  - — Document all of the investigation, analysis, and results when Root Cause Analysis is performed.
- *Developing and Implementing Corrective Action:* removal of the root cause of the problem or reduction of the occurrence of the root cause.
  - — Recognize that several corrective actions may be possible for each root cause.
  - — Evaluate candidates for effectiveness with respect to functionality and cost.
  - — Select the appropriate corrective action(s) and develop an implementation plan with assigned responsibility and scheduled milestones.
  - — Verify that actions have been implemented.
  - — Enter documentation into the R&M Data Collection System where it can be used to help prevent, diagnose, or mitigate future problems.
  - — Determine if original target has been addressed or needs modification based on new information.
- *Follow-up and Evaluation*
  - — Measure performance after implementation of corrective action.
  - — Compare to original data.
  - — Evaluate against original planned or modified target for improvement.

Root Cause Analysis and Failure Analysis

Root Cause Analysis is a logical, systematic approach to identifying the basic reasons (causes, mechanisms, etc.) for a problem, failure, non-conformance, process error, etc. Whenever a significant problem occurs (i.e., low frequency of occurrence with high cost in time and/or money, or high frequency of occurrence), Root Cause Analysis should be implemented. The result of Root Cause Analysis should always be the identification of the basic mechanism by which the problem occurs and a recommendation for corrective action. "No corrective action required" is an acceptable recommendation for corrective action when properly justified. Root Cause Analysis cannot be closed out until all required corrective action has been developed and implemented. Many users as well as professional societies (SAE, ASQC, IEEE, etc.) and the U.S. Government have developed methodologies for performing Root Cause Analysis. When Root Cause Analysis is performed, all of the investigation, analysis, and results (including corrective action) must be completely documented. This documentation should be entered into the R&M Data Collection System where it can be used to help prevent, diagnose, or mitigate future problems.

Failure Analysis is a special case in Root Cause Analysis where a physical failure has occurred. Failure Analysis is defined in the glossary and is synonymous with terms such as "Physics of Failure,"

"Reliability Physics," etc. Failure Analysis usually involves laboratory analysis and produces data that can include measurements of the failed item and analytical photographs using equipment such as X-ray machines, scanning electron microscopes, spectrographs, and optical microscopes. Proper preparation and preservation of samples and documentation of results can allow for independent expert evaluation of the analysis when disputes arise in determining the failure mechanism or the root cause of the failure mechanism.

Examples of disputes include:

- Defective part or part misapplication
- Overstressed part application or equipment operated outside of design range.

Like Root Cause Analysis, Failure Analysis is not complete until all recommended corrective actions have been developed and implemented.

Part II: Maintainability Tools and Techniques

Maintainability Assessment

Applying tools and techniques associated with maintainability starts as early as the Design and Development phase. Maintainability is designed into a system and does not come by chance. A Maintainability Assessment Tool such as the sample in Table E-1 can be used to evaluate the equipment maintainability during a Design Review, and when the equipment is in the process of being built. It is recommended that a cross-functional team from the user's organization perform this assessment.

Table E-1.Sample Maintainability Assessment Tool

Accessibility	Diagnosis	Skill Level	Tool Requirements	Alignment	Rating
Clear view/ Accessible	Highly detectable (indicates problem)	No training	No tools	Self	1
Clear view/ Inaccessible	Reasonably detectable	General skill level	Basic mechanic tools	Single point/ no tools required	3
Unclear view/ Inaccessible (special tools)	Moderately detectable (fault light)	Skilled trades	Trade-specific tools (micrometers, multimeter, etc.	Single point With tools	5
Unclear view Inaccessible (awkward body motions)	Difficult to detect (shows up in product/parts)	Special training	Special tools (tool room)	Multiple points/tools required	7
Unclear view/ Inaccessible (additional people/ equipment required)	Not detectable	Unique technology training	OEM patented tools/sophisticated test equipment	Multiple point by making parts	9

To apply this tool, the machinery or system must be evaluated by its subsystems, such as:

- Hydraulic systems
- Pneumatic systems
- Drive systems
- Control systems
- Application specific tooling, etc.

Each subsystem is then numerically rated independently for Accessibility, Diagnostics, Skill Level, etc. No totals are calculated. Every instance of a "7" or "9" rating should be targeted for improvement.

It is understood, however, that it may be necessary to accept scores higher than 5 when new technology is being introduced into the user's manufacturing operations. In this case, a plan to reduce these scores in the future should be generated.

Safety

Designing for safety and safe maintainability requires that safety engineering be introduced at the design stage. Safety personnel must be consulted up front to advise what appropriate safety measures must be integrated into the design.

Visual Management Techniques

Visual management techniques differ for varying types of equipment. The most appropriate visual management techniques can be determined best by using a team approach from both user and supplier. The techniques selected should be reviewed on a continual basis during concept and proposal, design and development, machine build, and on the manufacturing floor by all the team members. Examples of visual management techniques follow:

- Match marking of fasteners (nuts, bolts, screws, etc.) when appropriate
- Match marking of control adjustments (pressure, flow, temperature, speed, level, voltage, current, etc.) when appropriate
- Identification of normal operating ranges and levels
- Indication of direction of flow and product color coding on piping and hoses
- Indication of direction of rotation (drives, belts, chains, motors, etc.)
- Function labels (switches, valves, buttons, lights, etc.)
- Identification labels (cabinets, panels, boxes, etc.)
- Filters (lube, hydraulic and air) that indicate when dirty
- Filters labeled with replacement filter element number
- Belt and chain drives with guarding that permits quick visual inspection and access
- Replacement belt or chain number labels on guarding
- Labeling of each lube point with product number and color code

- Temperature-sensitive labels on all critical components (motors, drives, controls, hydraulic units, etc.)
- Equipment layout with all electrical control panel safety lockout points indicated (affixed to the main electrical control panel)
- Equipment layout with all lubrication fill points, frequencies, and product codes indicated (affixed to the main electrical control panel)
- Identification of all control drawing numbers on the main electrical control panel
- Signals or alarms that indicate a major abnormality, safety interlock tripped, process out of control, etc.
- Equipment and process inspection list (affixed to the main electrical control panel).

Appendix F
Figures of Merit

The following Figures of Merit are examples of R&M characteristics. It is important for the reader to understand that these Figures of Merit are not all-inclusive and that it may be necessary to develop special or unique measures for a specific application to address the production concerns of a user. The Figures of Merit shown in the following table were selected to offer the reader a range of possible measures and to provide a foundation for their appropriate use. It needs to be noted that appropriate measures must be selected and defined carefully to provide usable information to achieve R&M goals and not to become overly burdensome.

A supplier/user may develop a *special* Figure of Merit, for instance, to meet an internal or specialized need. An example of a *special* Figure of Merit that was developed internally by a body and assembly division of a large automaker is *Units lost per thousand*. This measure was defined as follows:

$$\textbf{Units lost per thousand} = \frac{\textbf{Number of units lost} \text{ (due to all or a particular causal factor)}}{\textbf{(Number of units produced} \div \textbf{1000)}}$$

This automaker could then apply this measure to all or specific causal factors to quantify and rank the causes of units lost.

Figure of Merit *Select Appropriate Measure*	R/M	Definition	Formula	Application	Notes (references are noted in brackets and are identified at the end of the table)
Achieved Availability (A_A)	R/M	Achieved Availability includes maintenance. It is a function of the Mean Time Between Maintenance Actions (MTBMA) and the Mean Maintenance Time (MMT).	$A_A = \frac{MTBMA}{(MTBMA + MMT)}$	Manufacturing systems	A measure to address production losses due to maintenance activities. MTMA is the arithmetic average of the time intervals between successive maintenance actions. MMT is the arithmetic average of the time required to perform maintenance. [Dodson et al.]
Availability, Intrinsic (A_I)	R/M	Operating time expressed as a percentage of uptime and repair time. See MTTR notes for definition of MTTR.	$A_I = \frac{MTBF}{MTBF + MTTR}$	Manufacturing systems	Intrinsic Availability is a measure of the Availability of a system ignoring all logistics issues such as obtaining personnel, material, and tools; and only considering uptime. Note that repair time refers to the time spent diagnosing, repairing the problem, verifying the repair, and restarting the equipment. Repair time does not include any logistics time. [Juran, MIL-HDBK-338].
Availability, Operational (A_O)	R/M	Operating time expressed as a percentage of operating and restore time. Operating time here can be gross operating time or uptime depending on how Availability is being calculated, that is, are we evaluating the functionality of the machinery or the production of parts.	$A_O = \frac{Operating\ Time}{(Operating\ Time + Restore\ Time)}$	Manufacturing systems	Operational availability is a more complete measure for system performance with respect to its operational goal of manufacturing, assembling, or testing product. Note that restore time refers to the time interval from when the equipment stops until it is back in operation. Restore time includes all logistics time. [Juran, MIL-HDBK-338].
Availability, Equipment (A_E)	R/M	The percentage of scheduled operating time (similar to uptime plus unscheduled downtime as defined in the Glossary) during which equipment is operable.	$A_E = \frac{Available\ Operating\ Time}{Scheduled\ Operating\ Time}$	Manufacturing systems	This measure was developed to isolate the equipment-related component of availability from an overall operational measure of availability. That is, influences due to facility, management, or process related interruptions are not included in the measure. Equipment Availability time definitions and method of calculation for times may be found in "Production Equipment Availability, A Measurement Guideline" [AMT].
b_{10} life	R	Life during which 10% of the population would have failed.	If the failure distribution for a component has been characterized, the b_{10} life can be determined from the probability density function at the 10% failed point. The value for the b_{10} life of components is usually provided by the manufacturer based on testing and analysis.	Mechanical components, Perishable tooling	This measure is calculated using probability distribution, but can be specified for any item, no matter what failure distribution is applicable. [Juran]
b_{50} life	R	Median life or life during which 50% of the population would have failed.	If the failure distribution for a component has been characterized, the b_{50} life can be determined from the probability density function at the 50% failed point. The value for the b_{50} life of components is usually provided by the manufacturer based on testing and analysis.	Mechanical components, Perishable tooling	This measure is calculated using probability distribution, but can be specified for any item, no matter what failure distribution is applicable. Note that the mean life (MTTF) can be greater or lesser than the median life. For the normal distribution they are equal, for the exponential distribution, median life is less than mean life. [Juran]
Durability Life (also called Time-to-wear-out or Longevity)	R	A measure of useable life defining the number of operating hours (or cycles) until overhaul or replacement is required. (Wear-out time for a machine)	Durability is not calculated, but is measured from installation until wear-out.	Manufacturing systems, Machinery, Perishable tooling	This is a specification value that is evaluated in operation. It is usually specified as a minimum value and helps define design parameters to ensure it is met. [Juran, Ireson]

Figure of Merit *Select Appropriate Measure*	R/M	Definition	Formula	Application	Notes (references are noted in brackets and are identified at the end of the table)
Failure Rate (λ)	R	Number of failures per unit of gross operating period in terms of time, events, cycles or number of parts.	$\lambda = \frac{\#\,failures}{Operating Time}$	Manufacturing systems	This is constant for the exponential distribution. For other distributions, it varies with operating time. Units of time in hundreds or thousands of hours or cycles produce numerically understandable results, i.e., 2 failures per 1000 operating hours. [Juran]
Mean Cycles Between Failure (MCBF)	R	Mean cycles between successive failures of a repairable product.	$MCBF = \frac{Operating\ time\ (in\ cycles)}{\#\,failures}$	Manufacturing systems	Same as MTBF, except uses cycles instead of time. See MTBF.
Mean Cycles to Restore (Repair) (MCTR)	M	The average cycles lost during restoration (repair) of machinery or equipment to specified conditions.	$MCTR = \frac{Total\ Restore\ (Repair)\ Time (in\ cycles)}{\#\,repairs}$	Manufacturing systems	Same as MTTR, except uses cycles instead of time. See MTTR.
Mean Down Time (MDT) or Mean Maintenance Time (MMT)	M	Arithmetic mean of down time per corrective maintenance action. Same as mean time to restore. Includes all unscheduled delay times.	$MDT = \frac{Total\ Restore\ Time}{\#\,repairs}$ $MMT = \frac{Total\ Maintenance\ Time}{\#\,actions}$	Manufacturing systems	Useful in assessing maintenance burden. [MIL-HDBK-338]
Mean Time Between Failures (MTBF)	R	Mean time between successive failures of a repairable product.	For data analysis: MTBF = Uptime (usually in hours) / Number of failures. For predictive analysis: $MTBF = \int_{t_0}^{\infty} R(t)dt$ where $R(t)$ is the reliability function for the system.	Manufacturing systems	MTBF is the inverse of failure rate for the exponential distribution. Failure rate produces results that can be less confusing to the non-reliability engineer. (Juran)
Mean Time Between Maintenance Actions (MTBMA or MTMA)	R	The arithmetic mean of time between maintenance actions.	$MTBMA = \frac{Total\ Operating\ Time}{\#\,maintenance\ actions}$	Manufacturing systems	This measure addresses all types of maintenance including preventive maintenance. Also referred to as Mean Time Between Maintenance (MTBM). [Juran]
Mean Time To Repair (MTTR)	M	The arithmetic mean of time to repair machinery or equipment to specified conditions.	$MTTR = \frac{Total\ Repair\ time}{\#\,repairs}$	Manufacturing systems, Control systems, Machinery	Confusion exists with respect to definition of MTTR. The "R" needs to be clearly defined. "Repair" time includes only time actually working on machine or control system, etc. "Restore" time includes repair time plus all delay times—waiting for personnel, parts, etc. It is preferable to use MTTR to address repair time, and Mean Down Time (MDT) to address restore time. [MIL-HDBK-338, Juran]

Figure of Merit *Select Appropriate Measure*	**R/M**	**Definition**	**Formula**	**Application**	**Notes (references are noted in brackets and are identified at the end of the table)**
Mean Time To Failure (MTTF) or Mean Life	R	Mean time to failure of a non-repairable product.	$MTTF = \dfrac{\text{Sum of the operating times}}{\#\text{failures}}$	Perishable tooling, Non-repairable components	This measure is used for non-repairable items. [Juran]
Overall Equipment Effectiveness (OEE)	R/M	The resultant of the combination of availability, performance efficiency, and yield factors such as the rate of quality products.	$OEE = \text{Availability} \times \text{performance efficiency} \times \text{yield}$	Manufacturing systems	A good measure of performance for evaluating changes to system over time. Difficult to predict because of interaction of manufacturing system with operating philosophy and part variability. Yield is the ratio of good parts produced to total parts produced. Performance efficiency is the product of the net operating rate times the operating speed rate.
Rate of Preventative Maintenance Actions (PMR)	M	Number of preventative maintenance actions per period of scheduled operating time or calendar time.	$PMR = \dfrac{\text{Preventive Maintenance Actions}}{\text{Scheduled (or Calendar) time}}$	Manufacturing systems	Useful in assessing maintenance burden. [Juran]
Reliability (R)	R	The probability that machinery/equipment/tooling can perform continuously, without failure, for a specified interval of time when operating under stated conditions.	Dependent upon Reliability Probability Distribution, for example: Exponential: $R = e^{-\lambda t}$ Weibull: $R = e^{\left[-\left(\frac{t-\gamma}{\eta}\right)^{\beta}\right]}$	Manufacturing systems, Electrical/electronic components, Mechanical components, Tooling	Theoretically, Reliability is the best measure of system performance; practically, it is a difficult value to calculate. Many distributions are time dependent, thus resulting in different answers at each point in the item's life. [MIL-HDBK-338]
Repairs/100	R	Number of repairs per 100 operating hours. This is usually uptime, but could be scheduled operating hours or net operating hours based on the customer's desire.	$\dfrac{\text{Number of Repairs}}{\left(\text{Operating hours}/100\right)}$	Manufacturing systems	Useful in assessing maintenance burden. [Juran]

References:

Dodson, Bryan and Nolan, Dennis, *Reliability Engineering Bible, The Complete Guide to the CRE*, Quality Publishing, Inc., 1995.

Juran, J. M., editor in chief, *Juran's Quality Control Handbook*, Fourth Edition, McGraw-Hill Inc., New York, 1988.

MIL-HDBK-338, Electronic Reliability Handbook, 15 October 1984.

AMT, *Production Equipment Availability – A Measurement Guideline*, The Association for Manufacturing Technology, McLean, VA, October 1998.

Ireson, W. Grant, editor in chief, *Reliability Handbook*, McGraw-Hill Inc., New York, 1966.

Appendix G

Data Tracking and Feedback System

If a data tracking and feedback system is included in the R&M plan, the supplier and user jointly determine what data to track, record and exchange and what, if any, defective parts are to be returned to the supplier. Although there are many differences between warranty and non-warranty actions, defective parts should be routinely tagged regardless of whether the warranty remains in force. It is very important that the data tracking and feedback system be consistently used if implemented, since the recording of accurate and meaningful data is critical to success of the system. A successfully implemented data tracking and feedback system will facilitate R&M growth and continuous improvement for current and future designs.

Figure G-1 depicts a tracking and feedback system. Figure G-2 provides the detail of a universal tag.

The R&M plan addresses recurrent failures and prescribes a process to resolve the design issues, manufacturing issues, and application issues concerning of a failing part. A supplier and user team may be formed to address the resolution of such issues.

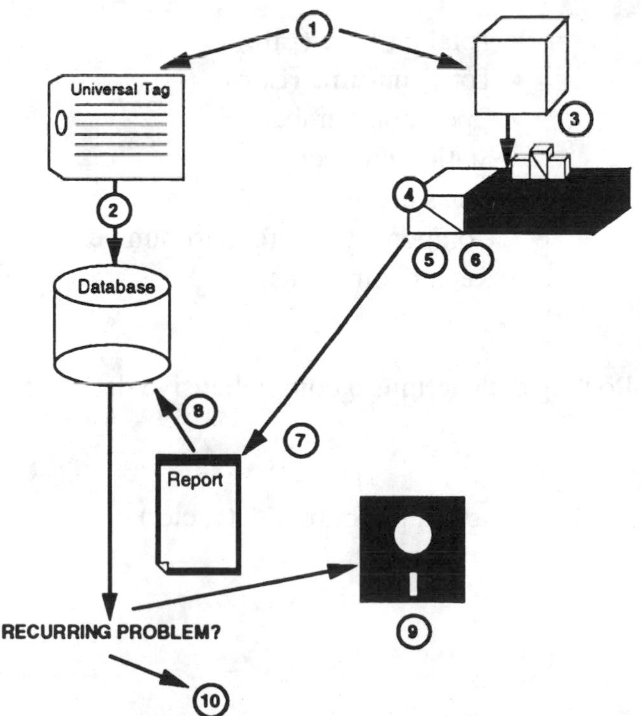

Process Steps:

1. A universal tag is attached to the failed component with a copy to be entered in the plant database.
2. Data is entered in the plant database.
3. A copy of the universal tag is sent to the manufacturing machinery and equipment supplier.
4. If appropriate, the component is sent to the component manufacturer.
5. An evaluation of the component is made to determine root cause of failure.
6. An identification of required corrective action is made.
7. A report is generated by the manufacturer and sent to the user and supplier.
8. The report is entered in the user and/or supplier databases.
9. A report containing all current information and reconciled activity is sent from the user database to the user's plant floor.
10. An exception report indicating recurring failures is generated from the user database. The user and / or supplier team takes action.

Figure G-1.Tracking and Feedback

Universal Tag

Must be completed when returning products for repair, regrind, exchange, warranty, rejection or misc.

Plant I.D. ______

Supplier/Manufacture ______ Date ______

Brass Tag No. ______ Bay No. ______ OEM ______

Op. No. ______ Sta. No. ______ Op. Name ______

W.O. ______ GL ______ SA ______ CC ______

Component Part No. ______ Serial No. ______

Description ______

Initial Observation ______

Fault Code ______ Time of Event ______

Time Event Cleared ______ Response Time ______

Repair Tme ______ Duration of Downtime ______

Total On-Time (life) ______ Total Run Time (life) ______

Work to be Done/Failure ______

Contact ______ Ext ______ Workunit ______

Did Replacement Solve Problem? Yes ____ No ____

Signature ______ Ext ______

Comments:

Repair ____ Exchange ____
Regrind ____ Warranty ____
Reject ____ Misc. ____
Attachements: Yes ____

PRESS HARD

FIRST COPY – GOES TO WORK UNIT (ORIGINAL)
SECOND COPY – GOES TO PLANT DATA BASE
THIRD COPY – GOES WITH THE PART TO THE MANUFACTURER

Figure G-2. Universal Tag

Data Feedback Requirements

In a tracking and feedback system basic data may be recorded by event, some examples follow:

- Plant/Location number
- Fault code
- Initial observation
- Time of event
- Time event cleared
- Response time
- Repair time
- Total downtime (total event time)
- Total on-time (Life)
- Total run-time (Life)
- Operation number
- Station number
- Brass tag number
- Component failure: part number
- Repair comments.

In addition to the above basic data, recording the following data permits a comprehensive database to be compiled:

- Description of root cause (if known)
- Machine state (such as proximity switch status, slide position, input/output data, etc.)
- Yield
- Performance efficiency.

Supplier Access to Data

The following paragraphs indicate several methods of providing R&M data to the supplier. These data may be available in electronic form or in paper form, as mutually agreed by user and supplier.

- **Periodic Supplier Plant Visits.** The user plant can allow access to its database by the supplier. Security issues have to be addressed to ensure accessibility. The supplier should take the initiative in collecting these data by making periodic visits to the plant and analyzing the data for set periods of time.
- **Local (Station) Data Dumps.** Event data can be captured and stored at the local level. Capture and storage of event data at the station or zone level makes it unnecessary for the supplier to access the plant monitoring database. The supplier is then allowed access to event data specific to the supplier's equipment any time supplier personnel are on-site.
- **Remote Access.** The machinery and equipment can be equipped to store event data at the local or zone level and to allow for secured remote data access by the supplier. The supplier can access these data at will via remote dial-up through a built in communication link to the machine controller.

Evaluation of Machine Performance (Delay Studies, Machinery Performance Studies, etc.)

To improve the baseline reliability of specific machinery and equipment, the user should encourage supplier visits and provide access to the plant to conduct short-term studies. These short-term studies should be joint efforts between supplier and user personnel.

Benefits of Plant Monitoring Systems and R&M Data in Plant Operations

- **Increased Uptime.** Critical to increased competitiveness is the ability to increase throughput with minimum cost increase. For this reason, plant management is very interested in obtaining improved uptime of equipment. R&M data are important aids to improve uptime.
- **Rapid Bottleneck Identification.** Production monitoring systems should have the ability to pinpoint production bottlenecks quickly. User management can then focus engineering resources on root cause analysis.
- **Improved Root Cause Analysis & Failure Analysis.** R&M data provide a basis for root cause analysis. Root cause analysis is needed for continuous improvement and improved equipment designs.
- **Continuous Improvement (CI) Monitoring—Reliability Growth.** Data feedback is essential for any continuous improvement program. It allows CI teams to focus on the strategically important areas, and provides quantifiable feedback of the impact of continuous improvement. Continuous improvement is a dual activity. It occurs within the user plant but also in the supplier organizations. Supplier access to the data drives a supplier's continuous improvement activity.

- **Reduced Life Cycle Cost (LCC).** The costs incurred during the Operation and Support phase contribute the largest segment of costs to LCC. R&M data collected during the Operation and Support phase provide a basis for monitoring and possibly reducing operation and support expenses by providing information necessary to identify problem areas that afford opportunities to reduce LCC.
- **Improved Production Equipment for Future Programs.** Continuous feedback from several user plants provides the supplier with vital information for improved designs for future programs. The data feedback can be used to update simulation models with real data for concept and design stages of new systems.

Sources of User R&M Data

Table G-1 lists possible sources of R&M data and the traditional and advanced methods of accessing the data.

Table G-1. R&M Data Sources

Data Source	Traditional Access Method	Advanced Access Method
Central Maintenance Records: • Dispatch system work order • Repair card system • Universal tag system • PM records • Supplier service records • Machine history	Plant Monitoring and Maintenance Feedback System: • Plant with existing machine tools – Typically 100% written manual/paper records • New plant and new machine tools – Automated, but not fully utilized – Maintenance history by paper records, some use of electronic data base systems but not linked to the plant monitoring system	Data entered or captured and stored in the machine controller. • Available for remote access upon demand • Maintenance history (corrective feed-back) data available and frequently used • Maintenance database systems designed to provide root cause analysis and summary data (made useful and easy to use)
Quality Control Charts (Predictor and Diagnostic tools for R&M)	Post or in-process gaging stores SPC data and provides screen presentation of chart data	SPC software has expert system to suggest possible root causes and/or assist in root cause analysis
Special Machine Tool Studies: • Delay studies • Laser alignment • High-speed photography/video Signature Analysis: • Vibration • Noise • Force/torque • Power consumption • Chemical analysis • Thermography	Supplier or user performs independently	Joint participation by supplier-user teams
Inventory Systems: • Spare parts consumption • Consumables (inserts, coolant, oils)	Inventory system not easily accessible to maintenance personnel	Computer inventory system accessible from the maintenance system
Equipment Monitor (excludes CNC) and Cell Controllers	PLC—limited failure event history captured	PC based—failure event history captured and stored for remote access. Auto cycle not allowed to resume until failure is documented either via control console or paper operator log/failure report.

Appendix H
Failure Mode and Effects Analysis (FMEA)

Introduction

Failure Mode and Effects Analysis is an effective qualitative tool in R&M practice and can be useful in any R&M program. Integration of this tool into the appropriate phase(s) of the R&M life cycle will set the foundation for continuous improvement and provide the basis for improved machinery, equipment, tooling, and process reliability and maintainability.

Purpose

The purpose of FMEA is to identify possible failure modes, determine the failure causes and effects for these failure modes, and assess the level of protection from these failure modes in the design and implementation of machinery, equipment, tooling, and processes. With the information gained from performing FMEA, R&M can be improved by modifying the design, the operation and/or the maintenance approach. These modifications can reduce the likelihood of a failure occurring, improve the detection and isolation of a failure, and/or reduce the downtime due to a failure. The FMEA uses an evaluation technique to help prioritize the implementation of recommended actions.

Note: If safety issues are identified during FMEA, they must be addressed in accordance with each company's safety procedures, and all safety analyses should be performed under qualified guidance and supervision.

Implementation

FMEA on user's manufacturing processes can be performed by the user during the development of the manufacturing process. This can occur prior to manufacturing and construction of the machinery and equipment during either the Concept and Proposal phase or the Design and Development phase, whichever is relevant.

FMEA on machinery can be integrated into the design and development process by the supplier to enable the results of the analysis to be considered prior to manufacturing and assembling the machinery and equipment.

FMEA on suppliers' internal processes can be performed to reduce the risk of introducing manufacturing defects that have a negative impact on machinery and equipment R&M.

FMEA is used to review a design or process, to minimize risk. As potential failure modes are identified, preventive action can be initiated to minimize or eliminate subsequent failures.

FMEA documentation can be used for communication to operating and maintenance personnel in user plants. This documentation provides value to the manufacturing operation by providing a basic troubleshooting guide. It can serve as a source document for maintenance logic diagrams and preventive maintenance plans. FMEA also provides a mechanism to document the rationale used for design or process changes.

FMEA Evaluation Criteria

A number of evaluation criteria have been published through various sources of FMEA material (DaimlerChrysler Corporation, Ford Motor Company, General Motors Corporation, Society of Automotive Engineers[2], Department of Defense, etc.). The purpose of evaluation criteria is to help establish priority for recommended actions resulting from the FMEA.

Guidelines for FMEA Application

A Process FMEA is used to analyze the manufacturing and assembly process. Potential product failure modes caused by manufacturing and assembly process deficiencies are identified.

A Design FMEA is used to analyze products, high-volume tools, perishable tooling, or machine components. These parts can be prototyped and tested before they are released to production. The Design FMEA focuses on potential failure modes caused by design deficiencies. This FMEA may provide inputs to the FMEA performed on machinery, equipment, tooling, process, and design.

A Machinery FMEA is used to analyze low-volume machinery and equipment where large-scale testing is impractical prior to production and manufacture of the machinery and equipment. The Machinery FMEA focuses on design changes to lower life cycle costs by improving the reliability and maintainability of the machinery and equipment.

The following examples are not all-inclusive, but offer some guidance in how the various types of FMEA are applied for different types of acquisitions. Note: There is no intent to prescribe an approach. It is the responsibility of each user and each supplier to understand the FMEA process and select the appropriate method, procedure, and items for FMEA in accordance with their business practice and contract requirements.

- Example: A cylinder block machining line is being acquired. The line consists of several transfer machines, each consisting of several machining stations, connecting automation, and gaging. A Process FMEA would be performed on the entire machining line and on each transfer machine. Machinery FMEA would be performed on individual machining stations (sometimes only on crucial operations), connecting automation, and in-line gaging. Design FMEA would be performed on the components and tooling.
- Example: A stand-alone grinding machine to grind a suspension component is being acquired. A Process FMEA would probably not be performed. Machinery FMEA would be performed on the grinding machine and its subassemblies. Design FMEA would be performed on the components and perishable tooling.

---

[2] The Society of Automotive Engineers' FMEA publication number J1739, *Potential Failure Mode and Effects Analysis in Design (Design FMEA) and Potential Failure Mode and Effects Analysis in Manufacturing and Assembly Processes (Process FMEA) Reference Manual* is a resource document of interest to manufacturing machinery and equipment users and suppliers in the ground vehicle industry.

Appendix I
R&M Training

Introduction

Training of supplier and user personnel is essential for the successful implementation of an R&M program for manufacturing machinery and equipment. Training should be sufficient to allow the company's employee to perform the job assigned.

Two categories of training are viewed to be important:

1. Awareness training in R&M basics
2. Detailed training in application of R&M practices and tools and techniques to specific tasks.

This training may be focused on the achievement of:

- An understanding of reliability and maintainability issues and how they impact business decisions and R&M improvement
- The use of the tools and techniques (such as FMEA, FTA, Figures of Merit, etc.) needed to assure that reliability and maintainability are designed in the manufacturing machinery and equipment
- An understanding of what, how, and when maintenance functions are to be conducted.

Training

Training may be a component of a company's human resource strategy. A commitment of resources to accomplish the training needs identified for the company's personnel should be part of a plan for training. Training should occur when it can be of most benefit to the employee—when the practices learned can be implemented immediately to increase retention and improve job performance.

Training can include various kinds of training media, including books, audio/visual tapes, self-paced instruction, formal coursework, consultant instruction, on the job training (OJT), etc. The type of training targeted should be appropriate to the employee's need(s), considering job requirements and the employee's proficiency or lack thereof. Figure I-1, Sample R&M Training Needs Assessment Form, is an example of an assessment form that a company may use to assess an employee's training needs relative to R&M. The form should be customized to suit the individual company's needs and the nature of its business.

Figure I-1. Sample R&M Training Needs Assessment Form

Employee Name:	Job Title:
Job Function:	Job R&M Proficiency Needed:
Assessment Date:	

Topic/Activity	Training* Accom-plished To Date	Training Needed**	Type of Training* Needed	State training objective(s)
R&M Basics				
Definition of R&M Requirements				
Stress Analysis and Design Margins				
Failure Modes and Effects Analysis				
Fault Tree Analysis				
Tolerance Studies				
Reliability Prediction Methodologies				
Other Figures of Merit Methodologies				
Maintainability Assessments				
Failure Analysis				
Machine Qualification				
Data Collection System Design				
Statistical Data Analysis				
LCC				

* Type of training:
A = Books, audio/visual tapes
B = Self-paced training program
C = Formal coursework
D = Consultant instruction
E = Degree program
F = OJT

** Level of training
0 = Not Required
A = Basic awareness
D = Detailed knowledge

Management Commitment

Management is responsible for evaluating the training needed by company personnel and for seeing that personnel receive awareness training or detailed training as appropriate. Management should oversee the integration of R&M into the daily job performance of company personnel.

- Supplier management needs to understand R&M principles and tools so it can direct the application of R&M in new and improved machinery and equipment designs and provide guidance and communication to its subordinates.
- User management needs to understand R&M principles and tools to provide direction and guidance to its subordinates and communicate the logic and justification for R&M requirements to purchasing and finance.

R&M Awareness Training

R&M techniques and practices cover many disciplines, ranging from statistical analyses to maintenance procedures. R&M awareness training allows an employee to gain sufficient familiarity with R&M principles, tools, and techniques to appropriately plan, manage, specify, or contract for new machinery and equipment and to monitor that the tools and techniques are being correctly applied.

R&M Detailed Training

Supplier Needs

Suppliers are concerned with providing educational experiences for their employees that include practices and techniques for designed-in R&M, and with providing quality instruction to the user's maintenance and operations personnel when necessary.

- Purchasing, sales, and engineering personnel need to understand R&M requirements in bid specifications, proposals, and sales contracts.
- Design engineering personnel are key to providing inherent R&M of machinery and equipment. What they design into the product greatly influences reliability, maintainability, and the life cycle cost to the user. Training in R&M for design personnel must be thorough and continuously improved. They must be proficient in the use of R&M tools and techniques, provide technical instruction, and compile and analyze data for continued maintenance and continuous improvement of the machinery and equipment.
- The machine builder must be familiar with the concepts of R&M in order to execute the design and to communicate any build-to-design issues back to the designer.
- Frequently, servicing personnel are the principal interface between the supplier and user after machine installation. They are responsible for demonstrating and communicating knowledge of R&M principles, such as selected Figures of Merit, data collection, maintainability issues, and root cause failure analysis, to the user and back to the designers. In the case of major systems, a service engineer invariably spends more time in the user plant than any other supplier representative and becomes familiar with what works and what does not work. Detailed training in R&M

principles will improve the quality and flow of information, resulting in continued improvement to machinery and equipment.

User Needs

User personnel should gain sufficient familiarity with R&M principles and tools to appropriately evaluate, specify, and contract for new machinery and equipment, and to correctly implement selected tools and techniques, e.g., Figures of Merit. Additionally, the training of operations and maintenance personnel in the care of user equipment throughout the machinery and equipment life cycle is necessary.

- Manufacturing engineering needs to understand reliability tools to appropriately specify the R&M requirements of new machinery and equipment. Manufacturing engineering along with the supplier's designers are the fulcrum of any R&M effort; and therefore, their training is critical.
- Machinery and equipment operators and maintenance personnel will realize the opportunity for R&M that is designed into the machine while they are operating and maintaining it. They may need operation-specific and/or maintenance-specific training to provide proper use and care of the machine. Proficiencies in data collection and failure analysis are needed also.

Glossary

Accelerated Life Testing

Testing to verify design reliability of machinery/equipment much sooner than if operating typically. This is intended especially for new technology, design changes, and ongoing development.[1]*

Acceptance Test (Qualification Test)

A test to determine machinery/equipment conformance to the qualification requirements in its equipment specifications.

Accessibility

The amount of available workspace around a component that is sufficient to diagnose, troubleshoot, and complete maintenance activities safely and effectively. Sufficient workspace allows for movement of necessary tools and equipment with consideration for human ergonomic limitations.

Actual Machine Cycle Time (Actual Cycle Time[2]) (Process Cycle Time)

Actual time to process a part or complete an operation. Specifically, the shortest time period at the end of which the sequence of events in the operation recurs.

Allocation

The process by which a top-level quantitative requirement is assigned to lower hardware items/ subsystems in relation to system-level reliability and maintainability goals.[3]

Availability

1. A percentage measure of the degree to which machinery and equipment are in an operable and committable state at the point in time when it is needed.
2. A measure of the degree to which tooling and equipment are in an operable and committable state at any point in time. Specifically, the percent of time that tooling and equipment will be operable when needed.[4]
3. The percentage of scheduled operating time during which equipment is operable, that is, operation is not prevented by equipment malfunction (or process difficulties for turnkey systems).[5]

See Appendix F, Figures of Merit.

Breakdown Maintenance

A policy whereby no maintenance is done unless and until an item no longer meets its functional standard.[6]

---

* Superscript numbers in this Glossary refer to Glossary References beginning on page 81.

Corrective Maintenance

1. Fixing things either when they are found to be failing or when they have failed.[7]
2. Maintenance done to bring an asset back to its standard functional performance.[6]
3. Tasks to correct existing problems.[7]

Cycle Time (Machine Cycle Time)

The period of time required by equipment or a system to perform one complete set of manufacturing operations on a product.[8]

Debug, Debugging

1. The operation of equipment or complex item prior to use to detect and replace components that are defective or expected to fail and to correct errors in fabrication or assembly.[9]
2. A process to detect and remedy inadequacies.[10]

Degraded, Degradation

A gradual impairment in ability to perform the required function.[9]

Design Machine Cycle Time (Process Cycle Time) (Ideal Cycle Time[2]) (Planned Cycle Time[5])

Specified time to process a part or complete an operation. Specifically, the shortest time period at the end of which the series of events in the operation recurs.

Detective Maintenance

Tasks designed to check whether something still works.[11]

Durability

Ability to perform intended function over a specified period under normal use with specified maintenance, without significant deterioration.[4]

Duty Cycle

1. The percentage of total power on time that an item operates.[1]
2. For electrical items, duty cycle equals 100% of the power on time the items perform foreground or background duties.[1]
3. For mechanical items, duty cycle is the percent of power on time the mechanics are in motion.[1]

Failure

An event when machinery/equipment is not available to produce parts at specified conditions when scheduled or is not capable of producing parts or performing scheduled operations to specification. For every failure, an action is required.

Failure Analysis (FA)

The logical, systematic examination of a failed item, its construction, application, and documentation to verify the reported failure, identify the failure mode, and determine the failure mechanism and its basic failure cause. To be adequate, the procedure must determine whether corrective action is warranted and if so, provide information to initiate corrective action.[1]

Failure Effect

The consequence of the failure.[1]

Failure Mode

The mechanism through which the failure occurs; also, the effect by which a failure is observed.[9] Generally, a failure mode describes the way the failure occurs and its impact on equipment operation.[12]

Failure Mode and Effects Analysis (FMEA)

A technique to identify each potential failure mode and its effect on machinery performance. See Appendix H.

Failure Rate

The total number of failures per unit of measure in terms of gross operating time, events, cycles, or number of parts, divided by the total number of units of measure, during a particular measurement interval.

Failure Reporting, Analysis, and Corrective Action System (FRACAS)

A process that provides for the orderly reporting, recording, transmission, and use of failure data. Failure data may include failure analyses, failure causes, corrective actions, etc.

Fault Tree Analysis (FTA)

A logic-based approach to failure analysis starting with an undesirable event and determining all the ways it can happen.

Gross Operating Time

Total time that the machine is powered and is actually producing parts.[13, 14]

Gross Operating Time = [Net Operating Time + Scrap Time + Machine Cycle Time]

Infant Mortality (Premature Failure)

Early failures that exist until debugging eliminates faulty components, improper assemblies, and other user and manufacturer learning problems.[1]

Launch Cost

The cost of starting production of a new product, or the cost of commissioning new equipment or a new system (includes the costs of site preparation, equipment installation, checkout and debugging, accreditation, and commissioning to specified levels of quality and productivity).[8]

Life Cycle

1. The sequence of phases through which machinery or equipment passes from conception through decommission.[15]
2. The period of time from the decision to design and install new equipment or a new system to its eventual retirement; or the period from the introduction of a new product until its withdrawal from the marketplace.[8]

Life Cycle Cost (LCC)

The total cost of ownership of machinery and equipment including its cost of acquisition, operation, maintenance, conversion and/or decommission.

Life Cycle Costing

A costing system that is based on an evaluation of all the factors that contribute to the cost of equipment or a system; or a product, over its life up to and including abandonment and salvage.[8]

Maintainability

A characteristic of design, installation and operation of machinery and equipment. Maintainability is usually expressed as the probability that a machine can be retained in or restored to specified operable condition within a specified interval of time, when maintenance is performed in accordance with prescribed procedures.

Maintenance

1. The upkeep of property or equipment.[16]
2. All actions necessary for retaining an item in or restoring it to a specified condition.[17]
3. Any activity carried out to retain an item in, or restore it to, an acceptable condition for use or to meet its functional standard.[6]
4. Work performed to maintain machinery and equipment in its original operating condition to the extent possible; includes scheduled and unscheduled maintenance but does not include minor construction or change work.[13]

See also **Breakdown Maintenance, Corrective Maintenance, Detective Maintenance, Planned Maintenance, Predictive Maintenance, Preventive Maintenance, Scheduled Maintenance, Unplanned Maintenance**, and **Unscheduled Maintenance**.

Modular, Modularity

The practice of designing and building equipment or a system in discrete blocks that can be re-arranged in a number of different configurations.[8]

Net Operating Time

Total time that the machine is producing first-time quality parts to specifications.

Non-operating Time

Total time that the machinery or equipment is up but not running due to blocked or starved conditions and/or administrative actions.

Overhaul

A comprehensive inspection and restoration of machinery/equipment, or one of its major parts, to an acceptable condition at a durability time or usage limit.[1]

Planned Maintenance

Maintenance carried out according to a documented plan of tasks, skills, and resources.[6]

Predictive Maintenance

1. Use of measured physical parameters against known engineering limits for detecting, analyzing, and correcting equipment problems before a failure occurs.[6]
2. A management technique that uses regular evaluation of the actual operating condition of plant equipment, production systems, and plant management functions to optimize total plant operation.[7]

Preventive Maintenance

1. Maintenance carried out at predetermined intervals, or to other prescribed criteria, and intended to reduce the likelihood of a functional failure.[6]
2. Overhauling items or replacing components at fixed intervals.[11]
3. A maintenance program that is committed to the elimination or prevention of corrective and breakdown maintenance tasks.[7]

Qualitative Measures

Criteria that are difficult, if at all possible, to measure[8] but that can be described or stated.

Quantitative Measures

Criteria that can be measured.

Reliability

The probability that machinery and equipment can perform [a stated function] continuously for a specified interval of time without failure, when operating under stated conditions.[18]

Reliability Growth

Improvement in machine reliability as a result of identifying and eliminating machinery or equipment failure causes during machine testing and operation.

Rule of Thumb

A practical guideline used to estimate values for an attribute or metric, as distinguished from an analytical or scientific approach.[8]

Scheduled [Planned] Downtime

The elapsed time that the machine is shut down for scheduled maintenance or is scheduled to be turned off for other reasons.[13]

Scheduled Maintenance

1. Any maintenance that is planned and prioritized to be done at a predetermined time.[6]
2. Any maintenance performed at a predetermined time, which is either planned or unplanned.

Testability

A design characteristic allowing the following to be determined with a given confidence and in specified time: location of any faults, whether an item is inoperable, operable but degraded, and/or operable.[1]

Time Elements

See chart below.

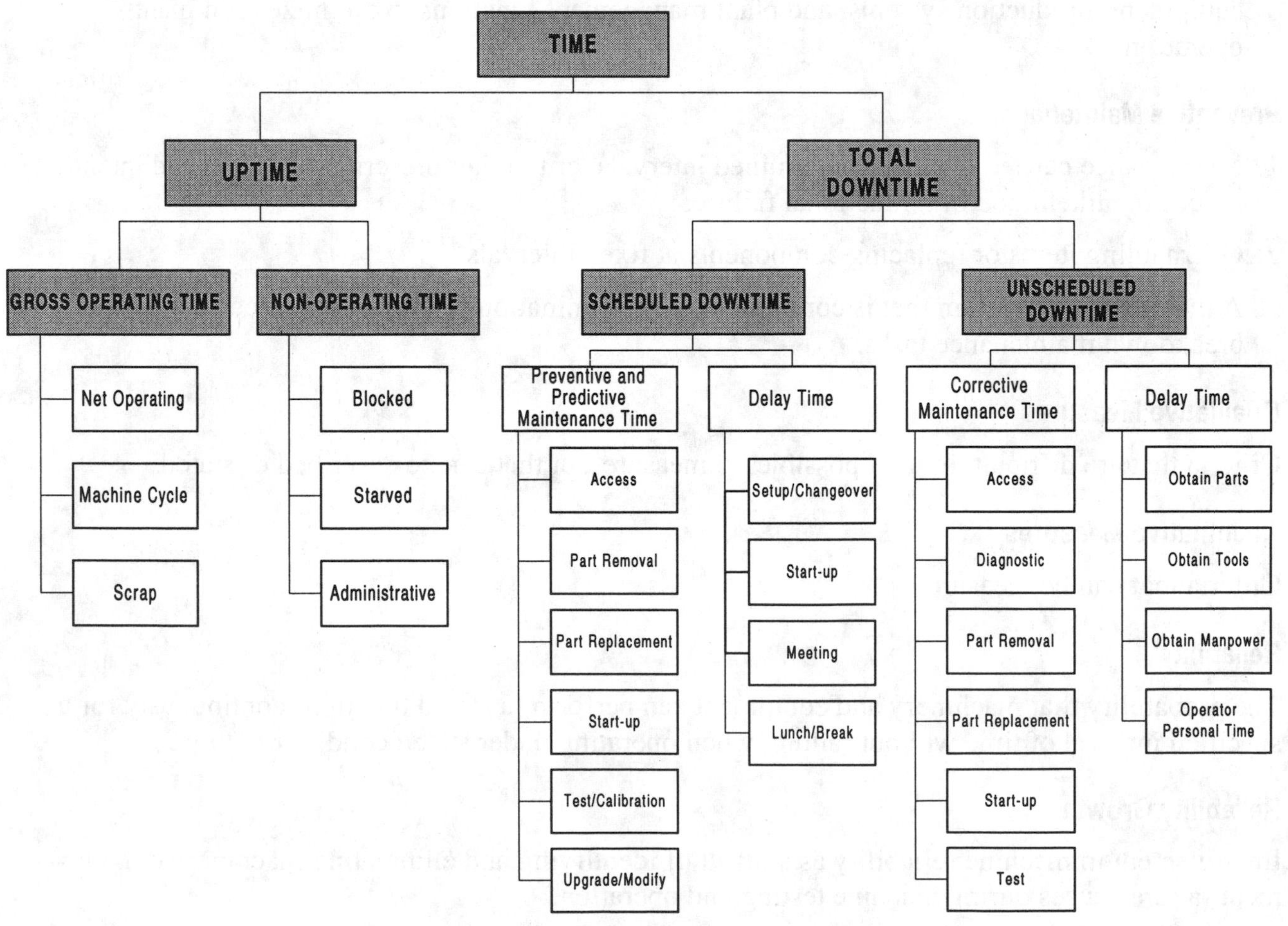

Sample Breakdown of Time Elements

Time to Repair (TTR)

Total clock time from the occurrence of failure of a component or system to the time when the component or system is restored to service (i.e., capable of producing good parts or performing operations within acceptable limits). Typical elements of repair time are: diagnostic time, troubleshooting time, waiting time for spare parts, replacement/fixing of broken parts, testing time, and restoring.[19]

Total Downtime

The elapsed time during which a machine is not capable of operating to specifications.

Total Downtime = [Scheduled Downtime + Unscheduled Downtime]

Unplanned Maintenance

1. Maintenance done without planning or scheduling; could be related to a breakdown, running repair, or corrective work.[6]
2. Maintenance done without planning. Typically, routine or minor maintenance where the cost of the planning exceeds any savings gained by a plan.

Unscheduled (Unplanned) Downtime

The elapsed time that the machine is incapable of operating to specifications because of unanticipated breakdowns.[13]

Unscheduled Maintenance

Any maintenance that is performed other than at a predetermined time.

Uptime

Total time that a machine is on-line (powered up) and capable of meeting functional requirements (e.g., capable of producing parts).[20]

Uptime = [Gross operating time + Non-operating time]

Yield

The percent of products meeting quality standards produced by the machinery or equipment.[2]

Glossary References

1. Omdahl, T. P. *Reliability, Availability, and Maintainability Dictionary*. ASQC Quality Press. 1988.

2. Nakajima, Seiichi. *Introduction to TPM, Total Productive Maintenance*. Productivity Press, Norwalk, CT. 1990.

3. MIL-STD-785. *Reliability Program for Systems and Equipment Development and Production.*

4. DaimlerChrysler Corporation, Ford Motor Company, and General Motors Corporation. *Quality System Requirements, Tooling and Equipment Supplement*. 1998.

5. The Association for Manufacturing Technology. *Production Equipment Availability, A Measurement Guideline*. AMT, McLean, VA. 1998.

6. Campbell, J. D. *Uptime: strategies for excellence in maintenance management*. Productivity Press, Portland, OR. 1995.

7. Higgins, L. R. *Maintenance Engineering Handbook*. 5th ed., McGraw-Hill Inc. 1995.

8. National Center for Manufacturing Sciences. *The Economic Evaluation of Flexible Manufacturing Systems*. Ann Arbor, MI. 1997.

9. Dummer, G. W. A. and R. C. Winton. *An Elementary Guide to Reliability*. 4th ed., Pergamon Press. 1990.

10. McKenna, T. and R. Oliverson. *Glossary of Reliability and Maintenance Terms*. Houston, TX: Gulf Pub. Co. 1997.

11. Moubray, J. *Maintenance Management – A New Paradigm*. Aladon Ltd. 1995.

12. MIL-STD-1629A. *Procedures for Performing a Failure Mode, Effects, and Criticality Analysis.*

13. General Motors Corporation. *GM Reliability and Maintainability Guideline for Machinery and Equipment*. October 1991.

14. ANSI/IEEE STD 500-1984 (R 1992). *Collection and Presentation of Electrical, Electronic and Sensing Component Reliability Data for Nuclear Power Generating Systems.*

15. MIL-STD-721C. *Definitions of Terms for Reliability and Maintainability.*

16. Merriam–Webster's Collegiate® Dictionary. 10th ed.

17. Dobson, B. and D. Nolan. *Reliability Engineering Bible*. Quality Publishing Inc. 1995.

18. Institute of Electrical and Electronic Engineers, Inc. *IEEE Recommended Practice for Design of Reliable Industrial and Commercial Power Systems*. Wiley-Interscience, New York, NY. 1980.

19. ANSI/IEEE STD 493. *Industrial and Commercial Power Systems*. 1980.

20. IEEE STD 100-1992. *Dictionary of Electrical and Electronics Terms.*

Selected Bibliography

A number of publications are available on the subjects of reliability and maintainability. The following selections were suggested by the contributors to this Guideline. To assist the reader in selecting reference and educational materials, they rated the complexity of the book selections:

I indicates suitability for gaining introductory-level or general information on reliability and/or maintainability.

E indicates suitability for gaining expert-level information on reliability and/or maintainability.

No rating indicates the material is rated somewhere between introductory and expert, and it may contain a range of levels of information.

Books

Abernathy, R. B. *The New Weibull Handbook*, 3rd ed., North Palm Beach, FL: R.B. Abernathy; Warrendale, PA: distributed by SAE, 1998.

Arsenault, J. E. and J. A. Roberts. *Reliability and Maintainability of Electronic Systems*, Rockville, MD: Computer Science Press, 1980.

Ascher, H. and H. Feingold. *Repairable Systems Reliability: Modeling, Inference, Misconceptions and Their Causes*, New York, NY: M. Dekker, 1984.

Association for Manufacturing Technology, *Production Equipment Availability, A Measurement Guideline*, McLean, VA, 1998.

Bennett, Robert E., J. A. Hendricks, D. E. Keys, and E. J. Rudnicki. *Cost Accounting for Factory Automation*, Montvale, NJ: National Association of Accountants, 1987.

Blanchard, B. S. *Logistics Engineering and Management*, 2nd ed. Englewood Cliffs, NJ: Prentice-Hall, Inc., 1981.

Bloch, H. P. and F. K. Geitner. *An Introduction to Machinery Reliability Assessment*, 2nd ed., Houston, TX: Gulf Pub. Co., 1994. **I**

Bloch, H. P. *Improving Machinery Reliability*, 3rd ed., Houston, TX: Gulf Pub. Co., 1998.

Bompas-Smith, J. H. *Mechanical Survival: The Use of Reliability Data*, New York, NY: McGraw-Hill, Inc., 1973.

Campbell, J. D. *Uptime: strategies for excellence in maintenance management*, Portland, OR: Productivity Press, 1995.

Canada, J. R. and W. G. Sullivan. *Economic and Multiattribute Evaluation of Advanced Manufacturing Systems*, Englewood Cliffs, NJ: Prentice-Hall, Inc., 1989.

Crowder, M. J. et al. *Statistical Analysis of Reliability Data*, 1st ed., New York, NY: Chapman & Hall, 1991. **E**

DaimlerChrysler Corporation, Ford Motor Company, and General Motors Corporation. *Quality System Requirements, Tooling and Equipment Supplement*, 1998.

Dobson, B. *Weibull Analysis*, Milwaukee, WI: ASQC Quality Press, 1994.

Dodson, B. and D. Nolan. *Reliability Engineering Bible, The Complete Guide to the CRE*, Quality Publishing, Inc., 1995. **I**

Doty, L. A. *Reliability for the Technologies*, New York, NY: Industrial Press, 1989. **I**

Dovich, R. A. *Reliability Statistics*, Milwaukee, WI: ASQC Press, 1990. **I**

Dummer, G. W. A., M. H. Tooley, and R. C. Winton. *An Elementary Guide to Reliability*, 5th ed., Oxford, England; Boston, MA: Butterworth-Heinemann, 1997. **I**

Ford Motor Company. *Reliability and Maintainability Guideline, The R&M Approach to Competitiveness,* 1990.

General Motors Corporation. *GM Reliability and Maintainability Guideline for Machinery and Equipment*, 1991.

Henley, E. J. and H. Kumamoto. *Probabilistic Risk Assessment: Reliability Engineering, Design, and Analysis*, New York, NY: IEEE Press, 1992.

Higgins, L. R. *Maintenance Engineering Handbook*, 5th ed., New York, NY: McGraw-Hill Inc., 1995.

Howell, R. A., J. D. Brown, S. R. Soucy, and A. H. Seed. *Management Accounting in the New Manufacturing Environment*, Montvale, NJ: National Association of Accountants, 1987.

Hoyland, A. and M. Rausand. *System Reliability Theory: Models and Statistical Methods*, New York, NY: Wiley, 1994. **E**

Institute of Electrical and Electronics Engineers. *IEEE Recommended Practice for Design of Reliable Industrial and Commercial Power Systems*, New York, NY: Wiley-Interscience, 1980.

Ireson, W. G., C. F. Coombs, Jr., and R. Y. Moss (eds). *Handbook of Reliability Engineering & Management*, 2nd ed., New York, NY: McGraw-Hill, Inc., 1996.

Juran, J. C. and F. M. Gryna. *Quality Planning and Analysis: From Product Development Through Use*, 3rd ed., New York, NY: McGraw-Hill, 1993.

Juran, J. C., editor-in-chief, A. B. Godfrey, co-editor-in-chief. *Juran's Quality Handbook*, 5th ed., New York, NY: McGraw-Hill, 1998.

Kapur, K. C. and L. R. Lamberson. *Reliability in Engineering Design*, New York, NY: John Wiley & Sons, Inc., 1977.

Kececioglu, D. *Reliability and Life Testing Handbook*, Englewood Cliffs, NJ: PTR Prentice-Hall, Inc., 1993.

Kececioglu, D. *Reliability Engineering Handbook*, Vol. I, II, Englewood Cliffs, NJ: Prentice-Hall, Inc., 1991.

Krishnamoorthi, K. S. *Reliability Methods for Engineers*, Milwaukee, WI: ASQC Press, 1992.

Kumamoto, H. and E. J. Henley. *Probabilistic Risk Assessment and Management for Engineers and Scientists*, 2nd ed., New York, NY: IEEE Press, 1996. **E**

Leemis, L. M. *Reliability: Probabilistic Models and Statistical Methods*, Englewood Cliffs, NJ: Prentice-Hall, 1995.

Lloyd, D. K. and M. Lipow. *Reliability: Management, Methods, and Mathematics*, 2nd ed., Redondo Beach, CA: Lloyd, 1977.

Locks, M. O. *Reliability, Maintainability, and Availability Assessment*, 2nd ed., Milwaukee, WI: ASQC Press, 1995. **I**

Mann, N. R. et al. *Methods for Statistical Analysis of Reliability and Life Data*, 1974. **E**

McKenna, T. and R. Oliverson. *Glossary of Reliability and Maintenance Terms*, Houston, TX: Gulf Pub. Co., 1997. **I**

Moubray, J. *Maintenance Management – A New Paradigm*, Aladon Ltd., 1995.

Moubray, J. *Reliability-Centered Maintenance*, 2nd ed., New York, NY: Industrial Press, 1997.

Nakajima, S. *Introduction to TPM, Total Productive Maintenance*, Cambridge, MA: Productivity Press, 1998.

National Center for Manufacturing Sciences. *The Economic Evaluation of Flexible Manufacturing Systems*, Ann Arbor, MI: NCMS, 1997. **I**

Nelson, W. *Accelerated Testing: Statistical Models, Test Plans and Data Analyses*, New York, NY: Wiley-Interscience, 1990. **E**

Nelson, W. *How to Analyze Reliability Data*, Milwaukee, WI: ASQC Press, 1983. **I**

O'Conner, P. D. T., D. Newton, and R. Bromley. *Practical Reliability Engineering*, 3rd ed., rev., Student Edition, J. Wiley, Chichester, NY, 1996. **I**

Omdahl, T. P., ed. *Reliability, Availability, and Maintainability (RAM) Dictionary*, Milwaukee, WI: ASQC Quality Press, 1988. **I**

Pecht, M., ed., and ARINC Research Corporation. *Product Reliability, Maintainability, and Supportability Handbook*, Boca Raton, FL: CRC Press, 1995.

Raheja, Dev G. *Assurance Technologies, Principles and Practices*, New York, NY: McGraw-Hill, Inc., 1991.

Shooman, M. L. *Probabilistic Reliability: An Engineering Approach*, 2nd ed., Malabar, FL: R. E. Krieger Pub. Co., 1990. **E**

Shooman, M. L. *Software Engineering: Design, Reliability, and Management*, New York, NY: McGraw-Hill, 1983. **E**

Society of Automotive Engineers. *Glossary of Reliability Terminology Associated with Automotive Electronics, SAE J1213/2*, Warrendale, PA: SAE, 1988.

Society of Automotive Engineers. *RMS, Reliability, Maintainability, and Supportability Guidebook*, 3rd ed., Warrendale, PA: SAE, 1995. **I**

Stamatis, D. H. *Failure Mode and Effect Analysis: FMEA from Theory to Execution*, Milwaukee, WI: ASQC Press, 1995. **I**

Wireman, T. *Total Productive Maintenance: An American Approach*, 1st ed., New York, NY: Industrial Press, 1991.

Industry and Military Standards

ANSI/IEEE STD 493-1990, *Recommended Practice for Design of Reliable Industrial and Commercial Power Systems (IEEE Gold Book).*

ANSI/IEEE STD 500-1984 (R 1992), *Collection and Presentation of Electrical, Electronic and Sensing Component Reliability Data for Nuclear Power Generating Systems.*

ANSI/SAE AIR 4276, *Survey Results: Computerization of Reliability, Maintainability, and Supportability (RM&S) in Design.*

IEC 60300-3-11 Ed. 1.0 b-1999, *Dependability management - Part 3-11: Application guide - Reliability centered maintenance.*

IEC 60605-1 Ed. 1.0 b-1978, *Equipment reliability testing. Part 1: General requirements.*

IEC 60605-2 Ed. 1.0 b-1994, *Equipment reliability testing - Part 2: Design of test cycles.*

NOTE: The above standards are a sampling of reliability standards and guides that can be obtained through the American National Standards Institute. Other reliability standards can be found on the ANSI web site.

IEEE STD-100-1996, *Standard Dictionary of Electrical and Electronic Terms.*

IEEE STD-1332-1998, *Reliability Program for the Development and Production of Electronic Systems and Equipment.*

IEEE STD-1413-1998, *Methodology for Reliability Prediction and Assessment for Electronic Systems and Equipment.*

SAE STD-J1739 July, 1994, *Potential Failure Mode and Effects Analysis in Design (Design FMEA) and Potential Failure Mode and Effects Analysis in Manufacturing and Assembly Processes (Process FMEA) Reference Manual*, Issuing Committee: Reliability/Maintainability / Supportability & Logistics

NOTE: The above standards are a sampling of reliability related standards available from the Institute for Electrical and Electronic Engineers and the Society of Automotive Engineers International. Other related standards can be found in their catalogues on the IEEE and SAE web sites. Other standard setting organizations, such as, American Society for Quality, Automotive Industry Action Group, etc. also publish standards that may relate to R&M. Their respective web sites and catalogues list their publications.

MIL-HDBK-217F, *Reliability Prediction of Electronic Equipment.*

MIL-HDBK-251, *Reliability/Design Thermal Applications.*

MIL-HDBK-263A, *Electrostatic Discharge Control Handbook for Protection of Electrical and Electronic Parts, Assemblies and Equipment (Excluding Electrically Initiated Explosive Devices)*

MIL-HDBK-338, *Electronic Reliability Design Handbook.*

MIL-HDBK-5F, *Metallic Materials and Elements for Aerospace Vehicle Structures.*

MIL-STD-1629A, *Procedures for Performing a Failure Mode, Effects, and Criticality Analysis.*

MIL-STD-2155, *Failure Reporting, Analysis and Corrective Action System (FRACAS).*

MIL-STD-721C, *Definitions of Terms for Reliability and Maintainability.*

MIL-STD-756B, *Reliability Modeling and Prediction.*

MIL-STD-785, *Reliability Program for Systems and Equipment Development and Production.*

NOTE: Military specifications and standards can be obtained from the National Technical Information Service. Ordering instructions are on the NTIS web site at http://www.ntis.gov/.

Periodicals and Newsletters

IEEE Transactions on Reliability, Reliability Society of the Institute of Electrical and Electronics Engineers, Inc. and ASQ Electronics Division. Quarterly publication.

International Journal of Quality & Reliability Management, MCB University. 9 issues per year covering reliability papers ranging from technical to managerial.

International Journal of Reliability, Quality, and Safety Engineering, World Scientific, River Edge, NJ. Quarterly publication.

Maintenance Technology, The Magazine of Plant Equipment Maintenance and Reliability Management, Applied Technology Publications, Inc., Barrington, IL.

MFPT Forum: Newsletter of the Society for Machinery Failure Prevention Technology, Society for Machinery Failure Prevention Technology, Haymarket, VA. Quarterly publication.

Reliability Engineering & System Safety, Elsevier Applied Science Publishers, Essex, England. Sixteen issues yearly. Formerly published as Reliability Engineering.

Reliability Magazine, Industrial Communications, Inc. Bimonthly periodical on reliability topics focused on plant equipment and maintenance.

Reliability Review, American Society for Quality Reliability Division. Quarterly periodical on general reliability topics.

Reliability Society Newsletter, Reliability Society of the Institute of Electrical and Electronics Engineers, Inc., New York, NY. Quarterly publication.

The RAC Journal, Reliability Analysis Center, Rome, NY. Quarterly periodical on general reliability topics.

NOTE: Journal, newsletter, and magazine availability is at the discretion of the publisher. Distribution may be limited to the organization's membership. Disclaimer: No endorsement is intended by this list. Any omissions are purely unintentional.

World Wide Web Sites

An extensive number of web sites can be found providing reliability and maintainability information. Major technical organizations as well as Government agencies have web sites. Government agencies and technical societies typically list their divisions and publications available. Some offer download capability from the web. For instance, a growing amount of R&M reference material can be found on the web from DOD and NASA. Use any search engine to search for the key words: "reliability" or "maintainability," or "reliability and maintainability." Using "maintainability" as the keyword was found generally to work best to find the type of informational sites being sought.